AI Nexus: Merging Intelligence, Security, and Networking

Gopalakrishna Karamchand

Satish Chitimoju

DEDICATION

This book is dedicated to my beloved family and parents, whose unwavering love, guidance, and sacrifices have been the foundation of everything I have achieved. Your constant support and belief in me have been my greatest source of strength, inspiring me to pursue my dreams with determination and purpose. The values you have instilled in me—curiosity, perseverance, and integrity—have shaped not only this work but also the person I am today. I am endlessly grateful for your encouragement, and it is with heartfelt gratitude that I dedicate this book to you.

CONTENTS

PREFACE

The integration of advanced artificial intelligence, networking, and security is becoming a scientific revolution for the new generation of communication systems across the globe. Today, networks are no longer dumb pipes that merely transmit information; they are complex entities capable of learning, self-organizing, and shielding themselves in real-time if need be. ***AI Nexus: Merging Intelligence, Security, and Networking*** delves into this fascinating intersection, exploring the technological advancements, challenges, and opportunities shaping the future of intelligent networks.

This book is a result of my personal interest in identifying how such technologies may revolutionise people's lives. It is meant to be used as the reference manual for people interested in understanding how networking systems using artificial intelligence work. That is why this work covers various aspects of the field, starting from the general definitions of the key concepts and potential discoveries to the concrete examples of the field's application and the discussion of the potential ethical issues.

Every effort has been made to ensure that this book is informative and practical, incorporating theoretical detail with real-world application. For anyone interested in this field as a researcher, a tech worker, or just as a layman walking the streets wondering about the future of intelligent systems, we hope this book has prepared you to engage with this field.

With the advent of today's third-wave AI as a pervasive component rather than a standalone object for global connectivity, let this journey unravel this phenomenon of the AI nexus. Together, let us build one that has networks that not only connect us but also reason, safeguard, and adapt to us.

ACKNOWLEDGEMENT

Writing ***AI Nexus: Merging Intelligence, Security, and Networking*** has been an extraordinary journey, and we owe a deep sense of gratitude to the many individuals who have supported and inspired me throughout this process.

First of all, we would like to thank my family for their support and understanding and for believing in me. We can only thank them for that because they are the foundation of my endeavours helping me to keep on going even when things get tough.

The authors and specialists who contributed their knowledge to the creation of important material are my thanks for helping me gain essential information about artificial intelligence, networking and cybersecurity. Lack of their help would make it difficult to provide meaningful and coherent information about the subject that forms the premise of this work.

All the same, a special acknowledgement to the academic and technical professionals whose research is shaping the future of this discipline. Several chapters were conceived owing to their endeavours, and this book is written with the modest intention of paying tribute to them.

We would also like to thank my editors, reviewers, and the entire publishing team. Without their helpful feedback and hard work, this manuscript could not have been turned into this work.

Now, this is something we am grateful for, you – the reader – for being interested in AI, security, and networking paradigms. Thus, it is my pleasure to dedicate this book to the cause of informing and empowering you to be an active participant in the future of intelligent systems.

Thank you all for being a part of this journey.

Chapter 01

THE FOUNDATION OF AI NETWORKING

Artificial intelligence, cybersecurity, and networking are among the most innovative forces in the modern interconnected world that revolutionize industries and change operational norms. Due to high capabilities of AI in manipulation, analysis and decision making on large data sets, network security and effective management of the networked infrastructure can be significantly improved. Across such areas as identification of advanced cyber threats to the enhancement of network functionality, AI has been the keystone in fostering organizations' ability to develop agile and high-impact systems that meets emergent technological environments. The blending of intelligence and infrastructure drives the idea of a shift to a new age where predictive power and automated means of reliability work harmoniously to protect assets.

AI is also used in networking and security, introducing a revolution of innovative solutions aimed at addressing emerging challenges. The conventional frameworks are filled up with technologies like machine learning-driven anomaly detection, use of predictive analytics to enhance the networks, usage of AI response mechanisms. When these domains are combined, one realizes not only more robust protection but improved scalability and flexibility of operation. These three aspects of intelligence, security, and networking play a role in creating a world of efficient communication and secure systems, so people and companies could succeed in the world of digital challenges.

The field of computer science known as AI focusses on programming computers to mimic human intelligence. A number of previously inconceivable tasks, including as mimicking human cognition and understanding speech, are now within the reach of researchers. In 1956, John McCarthy first used the phrase "AI" and described it as "the science and engineering of making intelligent machines." Scholars are attempting to dive more into the area of AI in order to provide answers for a variety of issues that include the involvement of machines, since AI has now become an integral component of the technology sector.

Computers may mimic human intellect via the use of AI, which is a method for developing software. At now, AI is finding usage in a wide range of applications, including intelligent gaming, processing of visual and auditory data, and natural language processing. Problems in AI, such natural language comprehension and generation, need a substantial quantity of information. AI makes use of predicate logic, which is useful for representing basic facts, to express knowledge. AI comes under several names, one of which is computational intelligence. AI has unquestionably contributed to improving people's quality of life, even if the advancements achieved in this area constitute just a small portion of the computer revolution.

1.1 Understanding Artificial Intelligence in Modern Systems

AI is the study of giving computers abilities that humans now possess. This transitory statement relates to the status of computer science at the present and leaves out important issue areas that are currently unsolvable by either humans or machines. Computer science and AI are very young disciplines. Work began in full swing immediately after WWII, and the term was first used in 1956. A lot of experts from other fields say that AI is the "field I would most like to be in" next to molecular biology. One may understandably assume that the likes of Galileo, Newton, Einstein, and the others had already claimed all the best ideas as a physics student. AI, however, is still looking for a few dedicated individuals who can fill the shoes of Einstein and Edison. Currently, AI covers a vast array of subfields, from the broad (perception and learning) to the narrow (proving mathematical theorems, producing poetry, controlling a vehicle through congested city streets, and illness diagnosis). It's true that AI is a discipline that applies to every intellectual endeavour.

A computer or system that mimics human intellect is at the heart of the AI concept. Machines imitating human behaviour and exhibiting human-like abilities such as perception, learning, predicting, reasoning, planning, and more are the ultimate aim of artificial intelligence research and development. The rapid advancement of industrial revolutions has led to the widespread usage of various sorts of machines, which in turn have rendered humans obsolete.

What is AI?

A key component of AI is the creation of representations and processes that can solve issues that people have traditionally addressed. The term AI has been defined differently by many writers. The majority of these explanations stay strictly on the scientific side, sidestepping any philosophical questions that may arise from seeing AI's ultimate goal as the creation of a synthetic human.

The following four groups make up these explanations:

1. Computing systems that mimic human intelligence by emphasising reasoning and a human-like framework.
2. Intelligent systems that are able to reason (with an emphasis on reasoning and intelligence in general).
3. Systems that mimic human conduct (with an emphasis on human structure and behaviour).
4. Intelligent systems that make logical decisions (with an emphasis on behaviour and intelligence in general).

AI is structured into three distinct levels: Artificial Intelligence, ML, and DL (Figure 1.1).

- AI was initially developed in the 1950s, allowing the computer to solve the issue by simulating human intellect.
- ML is a branch of AI that assists computers in learning and decision-making.
- DL is a subfield of ML that finds and categorises objects based on their characteristics extracted from massive datasets.

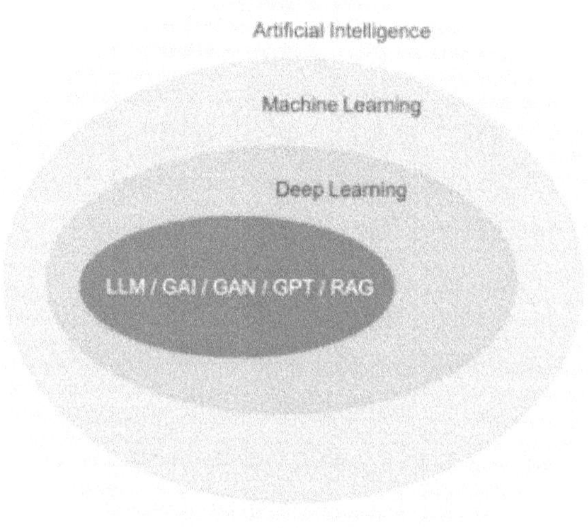

Figure 1.1: Overview of Artificial Intelligence

Source: - (D, 2023)

Currently, AI is being used extensively in many areas of our daily life, including healthcare, banking, retail, industry, agriculture, smart cities, and governance. The driver is guided by Tesla's autopilot system while changing lanes, driving through interchanges, and exiting highways. Autonomous vehicles and traffic sign recognition will soon be possible in the city. Google Home uses a NLP strategy to manage household appliances, play music, and respond to basic enquiries, in addition to its machine vision applications. In remote places, Amazon Prime Air' delivers tiny goods using drones. Both the shipping cost and the amount of time saved are significantly reduced.

Key Component of AI

In this section to discuss a key component of AI:

1. **Machine Leaning:** In the age of technology, we have enough of structured and unstructured data. Machine learning, a branch of AI developed in the late 20th century, makes predictions from data by use of self-learning algorithms. ML allows for the improvement of prediction models and data-driven judgements without the need for humans to manually derive rules and construct models from massive volumes of data. ML is playing an ever-larger role in both academic computer science and our everyday lives. Thanks to ML, we have trustworthy search engines, spam filters for email, software to recognise text and speech, and even hard chess systems. Autonomous vehicles that are both safe and efficient should be introduced soon. Researchers demonstrated that DL models can detect skin cancer with an accuracy close to that of humans. Researchers at DeepMind were the first to use DL to surpass physics-based approaches in predicting three-dimensional protein structures.

The three main branches of machine learning (ML)—supervised, unsupervised, and reinforcement learning—will be examined in this section. By comparing and contrasting the three main categories of learning, we may better grasp the real-world issue at hand, and we can do so by drawing on some conceptual examples. contexts in which they may be useful (Figure 1.2):

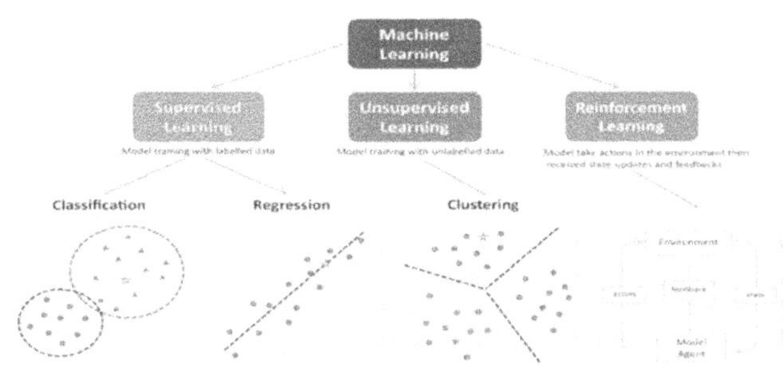

Figure 1.2: the fundamental three different types of machine learning

Source: - (Peng, 2021)

a. ***Supervised Learning.*** - supervised ML has shown to be a very effective and widely used sort of ML. In supervised learning, we utilise examples of input/output pairings to make predictions about future outcomes based on inputs. These input/output pairings make up our training set for the ML model. Classification and regression are the two main categories of supervised ML tasks.

b. ***Unsupervised Learning.*** - Unsupervised learning encompasses all forms of ML in which the learning algorithm is not taught by a teacher and there is no known outcome. Unsupervised learning involves just displaying the input data to the learning algorithm and asking it to draw conclusions from it. Unsupervised ML issues fall into two main categories: dataset modifications and clustering.

c. ***Reinforcement Learning:*** - Making choices to maximise cumulative rewards in a particular scenario is the focus of the machine learning field known as reinforcement learning (RL). In contrast to supervised learning, which uses a training dataset with predetermined responses, RL uses experience to teach. In RL, an agent learns to accomplish a task in an unpredictable, sometimes complicated environment by acting and getting feedback in the form of rewards or punishments.

2. **Deep learning:** DL is a subfield of ML. It is an algorithm that uses many processing layers with intricate structures or several nonlinear transformations in an effort to leverage the high-level abstraction of data. DL is an algorithm in ML that is based on characterising learning data. The idea of DL is related to learning that is done quickly. In the 1990s, shallow ML methods like LR and SVM were first presented. The numerous hidden layer nodes form the basis of DL. The multi-layer neural network is the foundation of DL. To learn extremely abstract data characteristics, DL employs the previous layer's input as the output of the subsequent layer.

3. **Natural Language Processing:** Humans speak natural languages. Programming languages like Java and C++ should be differentiated from this. Speech recognition and NLP are computer algorithms that allow computers to comprehend written and spoken human speech. This is linguistic intelligence, a particular kind of computerised pattern recognition.

    We have to distinguish the following types of application of natural language processing:

    - ***Speech-to-Text (STT)***: STT (Speech-to-Text) This program transforms spoken language into digital text. This is true when using Apple's Siri program, which dictates emails or notes straight into the device.
    - ***Speech-to-Speech (STS)***: Speech-to-Speech (STS) This kind of program may be found at Google Translate. This instantly converts an English voice input into a Chinese or Japanese speech output. For language output, the process known as natural language generation (NLG) is used. When using digital personal assistants (like Google Home or Alexa), this variation is also used in question-and-answer sessions. It should be referred to as STT—Processing—TTS more specifically. Since the digital assistants translate spoken language into digital text first, then understand and process it, create a digital text response, and have it read out linguistically—all in a matter of seconds!
    - ***Text-to-Speech (TTS)***: Text-to-Speech (TTS) This tool uses digital documents to generate a spoken rendition of the text. This allows you to "read aloud" emails, SMS, and other material. This also includes acoustic announcements in spoken dialogue systems. Those who are visually challenged may find this feature very useful since it allows them to "read" information on the screen.
    - ***Text-to-Text (TTT)***: Text-to-Text (TTT) A translation tool like DeepL or Google Translate is used in TTT applications

to translate one electronically accessible text into another language, likewise in text format.

4. **LLM (Large Language Model):** The task of large-scale human language modelling is very difficult and resource-intensive. It has taken many decades to enable language models and big language models to attain their present capabilities. Model complexity and effectiveness increase with increasing model size. A single word's likelihood could be predicted by early language models, but phrases, paragraphs, or even whole publications may now be predicted by massive language models. Over the last several years, as computer memory, dataset size, and processing capacity have increased and more efficient methods for modelling longer text sequences have been created, the size and capabilities of language models have grown. DL algorithms are used to train this kind of AI model, which can recognise, produce, translate, and/or summarise large amounts of textual data and written human language.

5. **GAI (Generative AI):** The ability of some AI systems to automatically produce new media (pictures, movies, text, etc.) in reaction to human input is known as generative artificial intelligence (GAI). These systems leverage generative models, particularly transformer-based DNN like LLMs, to learn patterns and structures from vast amounts of input data and then create new data with similar characteristics. GAI has gained significant attention with the advent of models like GPT-3and GPT-4, which are capable of producing coherent and contextually relevant text. These models use advanced mechanisms like "attention" to dynamically adjust the relevance of words in a given context, allowing for more accurate and efficient understanding and generation of content. The applications of GAI are diverse, ranging from chatbots and text-to-image generators to more complex systems capable of creating videos and music. Major tech companies such as OpenAI, Microsoft, Google, and others are at the forefront of developing and deploying these technologies.

6. **GAN (Generative Adversarial Network):** An architecture that falls under DL is a generative adversarial network (GAN). To generate the most genuine new data possible from a certain training dataset, it trains two NN to compete with one another. Use an existing picture database to create new images, or a song database to create new music. The adversarial nature of a GAN stems from the fact that it trains two separate networks in competition with one another. An example of how one network might produce new data is by applying extensive modifications to a sample of input data. It is the goal of the second network to determine whether the newly produced data indeed belongs in the first dataset. The predicting network, in other words, decides if the produced data is authentic or fraudulent. More and better false data values are produced by the system until the forecasting network is unable to tell the difference between the two. An algorithm that pits two neural networks—the discriminator and generator—against one another—thus the "adversarial" aspect. A zero-sum game is used for the contest, meaning that the success of one actor is the failure of another.

7. **GPT (Generative Pre-trained Transformer):** An important step forward in artificial intelligence, generative pre-trained transformers (GPTs) are a class of neural network models that enable generative AI apps like ChatGPT by using the transformer architecture. With GPT models, apps may generate content (pictures, music, and more) that sounds natural, and they can even answer questions like a person. Generative AI and GPT models are finding use in search, content creation, text summarisation, and question and answer bots across several sectors. Deep neural networks built on transformers that detect relationships in sequential input, such as words in a phrase, in order to understand meaning and context.

8. **RAG (Retrieval Augmented Generation):** RAG is an artificial intelligence framework that aids pre-trained big language models in producing more accurate, current information while decreasing hallucinations by retrieving data from an external knowledge

source. Retrieval Augmented Generation has garnered increasing attention due to its ability to overcome several limitations associated with traditional text generation models. However, generative models like as OpenAI's GPT struggle when faced with jobs that need accurate factual knowledge or intricate control over content, despite their impressive ability in producing coherent and contextually appropriate language. Integrates external knowledge sources to provide more accurate, up-to-date, and contextually relevant responses to GOT prompts.

1.2 The Evolution of Networking Technologies

Networking is so fundamental to modern life in this digital era. Networks have become indispensable in many aspects of modern life, from online banking to social media. In this article, we will explore the origins, development, current state, and potential future of networking. Since its start in the 1960s, it has gone a long way, and with the arrival of 5G, it now promises speeds that are lightning fast.

There have been few technical developments in the contemporary era as revolutionary as the development of networking. Today, networking plays an integral role in the way we communicate, conduct business, and connect. In this blog, we will explore the history, evolution, present stage, and future of networking, and highlight how networks are part of our daily lives.

Through networking, many computers may communicate with one another. The internet and other networking technologies will allow the two computers to be linked from different parts of the globe. More convenient and widely available nowadays are wireless modems, which eliminate the need for cables that connect to the computer system.

1.2.1 From Traditional to Modern Networking

Networking technologies development has advanced from physical interconnectivity solutions that have a low bandwidth and where most settings were configured manually to modern networking influenced by the internet, virtualization, and software-defined networks (SDN). The legacy networks had fixed designs and depended a lot on the physical devices while today's concepts focus on dynamism, expandability and smart control. From the onset, AI was integrated within networks using ML for traffic issues, abnormality detection, and self-healing routing.

Historical Overview of Networking Technologies:

Networking has advanced much from its infancy as a means of communication. Transmission of communications across great distances was made possible in the late 1800s by use of telegraph networks. The next big thing was the telephone, which broke new ground in communication by enabling instantaneous two-way conversations between individuals in different places. Networks of interconnected computers enabled distant data transfer in the middle of the twentieth century.

A new age of networking began with the introduction of the internet in the 1990s. The expansion of the World Wide Web and the advent of the internet enabled people all over the globe to communicate with one another and made previously inaccessible information and services readily available. The growth of social media platforms, mobile devices, and cloud computing has further accelerated the evolution of networking.

- **The First Computer Network is Born:** The first publicly accessible computer network, ARPANET (Advanced Research Projects Agency Network), went live in 1969, marking the beginning of the current era of computer networking technologies. As the Internet evolved from its TCP/IP protocol stack, it orchestrated its implementation. The Advanced Research Projects Agency (ARPA), which is a division of the United States Department of Defence, came up with ARPANET. By replacing direct connections with packet-switching, ARPANET completely changed the face of communications. For data transmission over a packet-switching system, each packet is pre-formatted with the address of the receiving computer before being sent out into the network.

- **Attached Resource Computer Network (ARCNET):** Datapoint Corporation first created the ARCNET communications protocol for LANs in 1986. In the 1980s, it was extensively used for office automation, and it was the first networking system that was widely accessible. For the first time, ARCNET did not presume any particular kind of computer networking technology would be linked, in contrast to previous computer systems that mandated that all networked machines be identical.

- **Emergence of Wireless and High-Speed Internet:** Wi-Fi plays a crucial and priceless role in modern media activities. We no longer separate wireless networks from our media ecosystem; they are now as essential to our daily lives as running water or electricity. Almost every house in the industrialised world now has its very own wireless internet connection. Worldwide, 169 million public Wi-Fi hotspots were available in 2018. The number of public Wi-Fi hotspots is expected to reach 628 million by 2023. There are new business and residential buildings going built. We'll begin by keeping our discussion general enough to cover a wide range of networks, including both wireless LANs such as WiFi and 4G and 5G cellular networks; we'll drill down into a more detailed discussion of specific wireless architectures in later sections.

1. **Wireless hosts:** Client devices that execute programs are known as hosts in wireless networks, just as they are in wired ones. The term "wireless host" may refer to a wide variety of devices, including mobile phones, tablets, laptops, and even sensors, appliances, cars, and other IoT items.

2. **Wireless links:** Hosts establish wireless connections with base stations (to be described later) or other wireless hosts. Both the transmission rate and the maximum distance that various wireless connection technologies are capable of covering are not constant.

3. **Base station:** An essential component of every wireless network is the base station. In a wired network, a base station is not immediately comparable to a wireless host or wireless connection. A base station's primary function is to communicate with wireless hosts via the transmission and reception of data packets.

4. **WiFi:** 802.11 Wireless LANs have quickly become one of the most crucial Internet access technologies, found everywhere from homes and schools to cafes, airports, and even street corners. While other wireless LAN protocols and technologies were created in the 1990s, the IEEE 802.11 wireless LAN, more often known as WiFi, has stood out as the obvious victor. A detailed analysis of 802.11 wireless LANs, including their frame structure, medium access protocol, and interworking with wired Ethernet LANs, will be conducted in this part.

1.2.2 Milestones in Networking Technology Development

The creation of ARPANET (1969), the development of TCP/IP protocol suite (1970s), the emergence of Ethernet (1970s), the introduction of the DNS (1980s), the birth of the WWW (late 1980s), a commercialization of an internet (1990s), the rise of broadband and social media (2000s), and the adoption of mobile internet and cloud computing (2010s). In this section to discuss a milestone, key technologies breakthroughs in networking, history of networking.

Overview of key milestones and developments:1960s: The Precursors

1. **ARPANET:**

   The ARPANET, which was established in the 1960s and supported by the U.S. Department of Defence, is considered the progenitor of the modern Internet. The original aim of ARPANET was to provide a means of communication between various government organisations and academic institutions.

2. **TCP/IP and Email 1970s:**

The goal of developing the TCP and the Internet Protocol (IP) was to provide standards for information exchange across various networks. This paved the way for the creation of the modern internet. Ray Tomlinson sent the first email in 1971, providing a new means of communication.

3. **Domain Name System (DNS) and World Wide Web 1980s:**

In 1983, the DNS was developed to convert domain names into IP addresses in a way that humans can understand. Tim Berners-Lee, a British scientist, developed the WWW in 1989, introducing the concept of websites and URLs.

4. **Commercialization and the Dot-Com Boom 1990s:**

The internet transitioned from a research and academic tool to a platform for commerce and communication. In 1994, the need for web standards led to the establishment of the WWW Consortium (W3C). There was a proliferation of internet-based businesses that witnessed fast expansion and investment during the dot-com boom that began in the mid-1990s.

5. **Broadband and Social Media 2000s:**

The proliferation of broadband internet enabled more consistent and quicker connections. Friendster, Myspace, Facebook, Twitter, and LinkedIn were all once-popular social networking sites that revolutionised the way people communicated and shared information.

6. **Mobile Internet and Cloud Computing 2010s:**

The proliferation of smartphones and other portable electronic gadgets has resulted in a surge in mobile internet use. The ability to access and store data over the internet, or "the cloud," quickly became the ruling paradigm in computing. Everyday objects were linked to the internet via the emergence of the IoT.

7. **Continued Evolution 2020s:**

   Technological developments such as 5G, AI, and blockchain are constantly changing the internet. The need to address privacy and security issues has sparked debates on the regulation and oversight of the internet.

Key technological breakthroughs in networking, such as Ethernet, Wi-Fi, and 5G.

When many computers are linked together, it is called networking. Sharing information between members is the main function of a network. Connecting more than two computers to a network necessitates the use of a hub or port, as seen in figure 1.7. A few examples of cellular network technologies include Ethernet, which allows for fast and dependable connections in wired LANs, Wi-Fi, which allows for wireless LANs and convenient connectivity, and 5G, the most recent generation of cellular network technology, which enables new applications like autonomous vehicles and augmented reality with much faster speeds and lower latency than previous generations. Breakdown of every technology.

Ethernet: Ethernet has quickly become the standard for LANs in both homes and businesses. Not only is it inexpensive, but it's also quick and simple to set up. A built-in Ethernet connectivity is now standard on most computerised devices. Given these considerations, it's no surprise that some industrial control system makers are pushing for Ethernet—with a few tweaks here and there—to facilitate real-time communications on the production floor. These suggestions are sometimes referred to as "Industrial Ethernet" despite the fact that they represent distinct and, frequently, conflicting answers. The backbone of local area networks is Ethernet. The level of Ethernet standardisation in the contemporary LAN market is uncertain. Even though it has a few drawbacks, Ethernet dominates all other technologies because of its massive market share (figure 1.3).

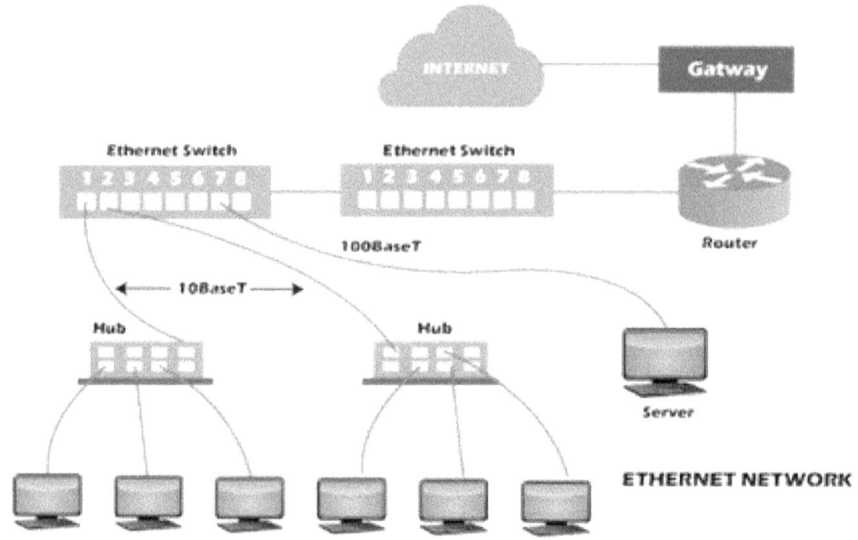

Figure 1.3: Overview of Ethernet network
Source: - (Falcon, 2024)

A short historical overview:

- The 2.94 Mbps version of Ethernet was first created in the 1970s by the Palo Alto Research Centre (PARC) of Xerox. An industry group consisting of DEC, Intel, and Xerox standardised the technology as Ethernet Version 1 (also known as DIX V1.0 standard). This version increased Ethernet's speed to 10 Mbps while preserving Ethernet's heavy trunk cable structure. DIX V2.0, more often known as Ethernet II, was published in 1982. Afterwards, Xerox gave up its trademarks.

- The IEEE 802.3 standard, which is a formal definition of Ethernet, was issued by the Institute of Electrical and Electronic Engineers (IEEE) in the mid-1980s. Token Ring and other competing local area network (LAN) technologies were mostly replaced by the original 802.3 Ethernet, which ran at 10Mbps.

- An article titled "IEEE 802.3 Carrier Sense Multiple Access with Collision Detection (CSMA/CD) Access Method and Physical Layer Specifications" was released in 1985 by the IEEE 802.3

committee. Although the 802.3 CSMA/CD standard varies from DIX V2.0 in terms of frame specification, the technology is more often known as Ethernet. There is no guarantee of compatibility between 802.3 and DIX frames, even if they may coexist on the same cable. Thus, 802.3 frames or DIX V2.0 frames must be defined when the topic of "Ethernet" is brought up.

- In 1983: The original IEEE standard for Ethernet technology is released by the IEEE. Under the moniker IEEE802.3 CarrierSenseMultipleAccess with CollisionDetectionAccessMethod and Physical Layer Specifications, it was created by the 802.3 subcommittee of the IEEE802 committee. The frame determination definition was one area where IEEE revised the DIX standard.

<u>Wi-Fi Technology:</u>

Wi-Fi standards may be categorised into four groups: 802.11b, 802.11a, 802.11g, and 802.11n, with each group corresponding to a different speed and presentation time. Commonly used standards are IEEE802.11b and IEEE802.11g [2]. One of the most popular Wi-Fi standards, IEEE802.11b has been around longer than any other wireless network standard. The highest speed it can achieve is 11 Mbps. A bandwidth of 5.5 Mbps, 2 Mbps, or 1 Mbps may be selected for use in situations where the signal is weak or interferences are present. Maintaining a steady and dependable network is made possible by automatically conditioning bandwidth.

The throughput of IEEE 802.11a is higher than that of IEEE 802.11b. With strong anti-interference capabilities, it operates in the 5.8GHz frequency range. However, neither IEEE 802.11b nor IEEE 802.11g will work with it. Also, it only covers about 30 square meters inside, which isn't very big. Thus, out of all the Wi-Fi technologies available today, IEEE 802.11a is by no means the most popular.

The IEEE 802.11g standard was officially ratified in July 2003[2] to address the issues with IEEE 802.11a and IEEE 802.11b, which were incompatible with one another. Conformity with IEEE 802.11b is a possibility. Therefore, there are more uses for IEEE 802.11g than for IEEE 802.11a.

The most recent Wi-Fi standard was ratified by IEEE in 2009 and is known as 802.11n. The transmission speed may reach up to 600 Mbps, with a norm of 300 Mbps.

One of the most fundamental aspects of Wi-Fi is the transmission of radio signals. Embedded systems engineers may get several advantages from wireless rabbits in various areas. Figure 1.4 illustrates the Rabbit's role in a sensor monitoring application.

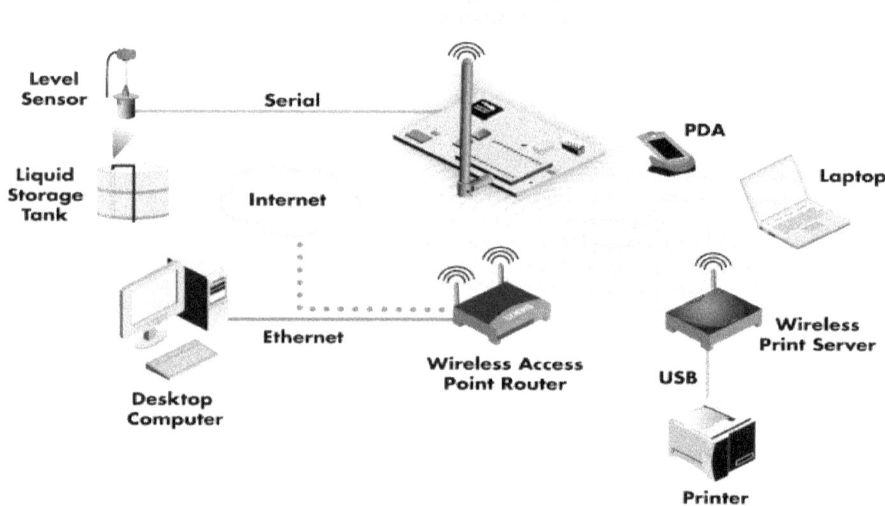

Figure 1.4: Wireless Local Area Network Connected to the Internet
Source: - (Perets, 2019)

5G Technology: Early 1970s saw the beginning of wireless communication. From 1G to 5G, mobile wireless technology has come a long way in the last 40 years. "5G" refers to the latest release in wireless technology. It provides users with unprecedented bandwidth. The most powerful and in-demand technology of the future will be 5th generation because of all the new sophisticated features it provides.

The acronym for "5th Generation Mobile" describes this kind of mobile network. With the advent of 5G technology, mobile phone use inside very high bandwidth has been revolutionised. This level of high-value technology has never been encountered by the user before. Cell phone (mobile) technology is now well known among mobile users. The inclusion of all sorts of cutting-edge features in 5G networks makes them the most powerful and most sought-after networks of the future.

The sheer magnitude of the cutting-edge features packed into brand-new mobile phones is mind-boggling. The 5G technology available on smartphones now outperforms a thousand lunar modules in terms of both power and feature set. An additional option for those looking for broadband internet access is to connect their 5G mobile phone to a laptop. Imagine a world where 5G technology powers your camera, MP3 recorder, video player, massive phone memory, lightning-fast dialling, audio player, and many more unexpected features. Bluetooth and Pico nets have exploded in popularity as a kind of kid-friendly rocking entertainment. In Fig. 1.5, features of 5G are shown that more elaborate necessity of 5G technologies.

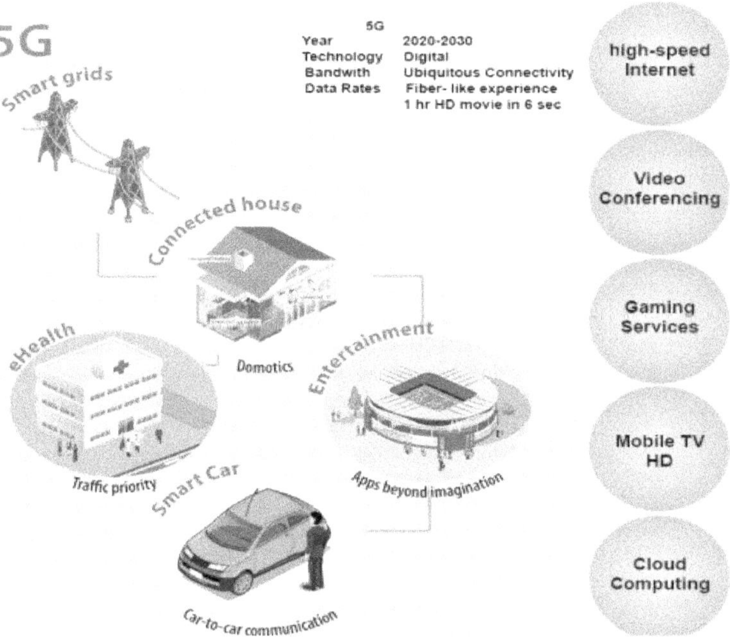

Figure 1.5: 5G-Wireless Communication Technology
Source: - (Fizza & Shah, 2016)

Types of Communication Network:

1. Local Area Network:

The term "local area network" (LAN) refers to a kind of computer network that links computers in a specific physical location, such as a home, school, lab, university campus, or workplace. In terms of LAN technology, the two most popular options are Ethernet and Wi-Fi. In the 1970s, researchers created a slew of LAN technologies that would later become commercially available. In 1973 and 1974, Xerox PARC worked on developing Ethernet.

Users on a wireless LAN are free to go wherever the signal is strong enough. Because they are so easy to set up, wireless networks have recently exploded in popularity among homes and small companies. Since Wi-Fi is already integrated into most mobile devices, most wireless LANs make use of it. A hotspot service often provides guests with Internet connection.

LANs were born out of this. By eliminating the need for a central hub for data interchange, these networks made it possible for computers to work together on computing tasks, hence facilitating the decentralisation of computing.

The industrial arena was the first to adopt local area networks; before, a centralised PLC or main control system had been used for control. Many PLCs, each with its own intelligence to execute control, might be located in various places thanks to LANs. They also made it possible for PLCs to share system data with one another and with other PLCs in the plant that were responsible for controlling other processes. I/O bus networks are a subset of networking that emerged with the industrial revolution; they enable smart field devices to talk to PLCs even when they don't have the usual input/output connectors. The next chapter provides a detailed explanation of I/O bus networks (figure 1.6).

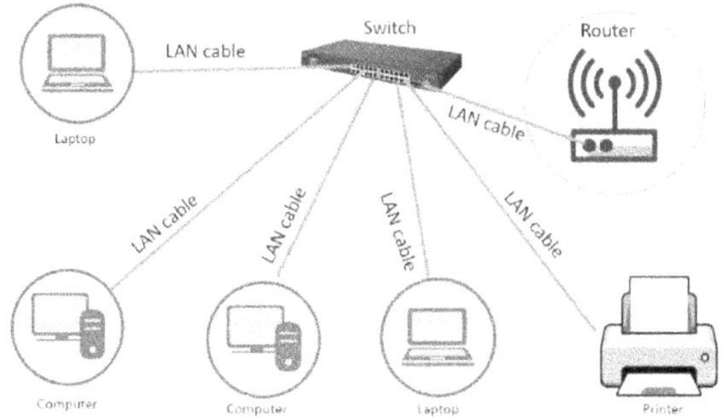

Local Area Network

Figure 1.6: structure of LAN

Source: - (Rehman, 2021)

2. Wide Area Network:

WAN links users across greater distances. Interstate, intrastate, and international connections are its primary use cases. WAN are shown in figure 1.7. Data transmission in WANs often occurs via the public telephone network and satellite connections. Worldwide area networks often use data transfer speeds lower than 1 Mbps. Typically, it is held by a number of different entities. Because of the larger distances and varied transmission media employed, the transmission time for WAN is higher.

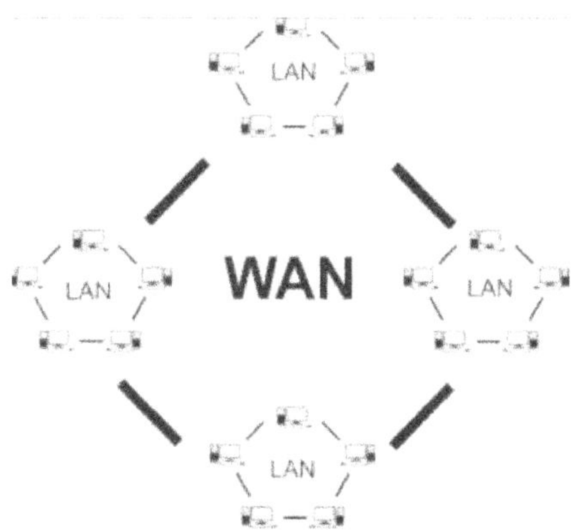

Figure 1.7: Structure of WAN

Source: - (Ambedkar, 2015)

3. Metropolitan Area Network (MAN):

Compared to LAN, it requires a lot more space, but compared to WAN, it's smaller. What we have here is an individual-owned and -operated network of computers. Since both enterprise-level high-speed LANs and WANs were not yet widely available, the necessity for a very fast networking capability prompted the development of MANs. WANs include switches and common carrier lines. These circuits originally connected a terminal to a distant computer via low-speed voice-grade communications channels. Optical and then digital circuits allowed for high-speed connections between distant computers, eventually replacing voice grade circuits (figure 1.8).

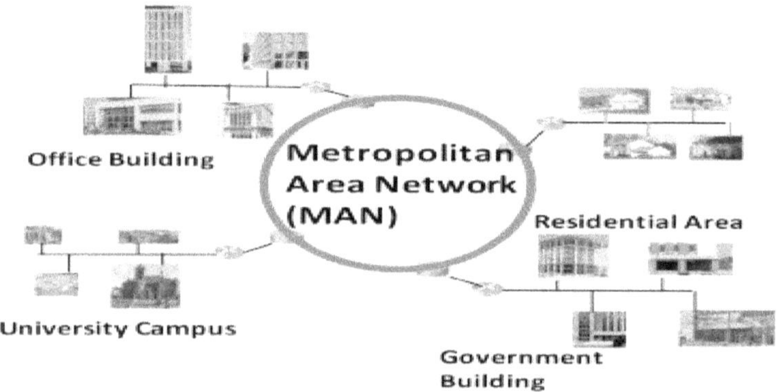

Figure 1.8: Structure of MAN

Source: - (Majumdar, 2019)

<u>Components of a Local Area Network: -</u>

The three main components of any data communication network are the following: a transmission medium, which can be twisted pair, coaxial cable, or fibre optics; a mechanism to control the transmission over the medium; and an interface for the host computers or other devices, called nodes, that are linked to the network.

- **Network Devices**

  - **Switches:** Increases the quantity of Ethernet ports that may be used for network connections. Direct data to its rightful destination, manage data flow, and set up special routes.

- **Routers:** Device in the heart of a network, responsible for controlling and directing data transfers between nodes. Find the optimal data route, connect several LANs or segments.
- **Hubs:** Standard networking hardware that establishes connections between various nodes in a network
- **Access Points**: Wi-Fi allows for the activation of wireless LANs.

Ethernet Cabling Types:

Three distinct kinds of cabling are compatible with Ethernet deployments. As a shown in Figure 1.8:

- **Coaxial cable: -**

  The components of coaxial cable, which is sometimes shortened to "coax," are a single conductor, insulation, a metal shield, and a plastic sheath. A signal's attenuation—a weakening of its intensity and quality—can be caused by electromagnetic interference (EMI), which the shield helps to prevent. EMI may be produced by several devices, including microwaves, mobile phones, radio transmitters, and fluorescent lamp ballasts.

- **Twisted-pair cabling:**

  Industrial point-to-point applications using twisted-pair wires may achieve transmission speeds of up to 250 kilobaud across lengths of up to 4000 feet. In addition to having decent noise immunity (which is enhanced when insulated), twisted-pair conductors are cheap. Addition of nodes to a twisted-pair bus, however, causes a precipitous decline in performance. Even more so, these conductors' performance is impaired by nonuniformity. Because the cable's characteristic impedance changes at different points, there is no one "correct" value for termination resistance, which makes it hard to decrease reflections.

- **Fibre optic cabling:**

Tightly packed into a low-refraction medium are tiny glass or plastic fibres that make up fibre-optic cable. Pulses of reflected light are used to carry signals over this kind of wire. There is currently no ideal low loss terminal access point (tap, T-connector, etc.) for fibre optics, which is the technology's biggest drawback. Currently, the T-connectors used in fiber-optic cables only absorb a fraction of the light energy that travels through the channel. This shortcoming renders fibre optics useless in big bus topologies, however they may still be used in star or ring topologies. Aside from that, optical couplers are much more costly than purely electrical connections, and fiber-optic cable is three to four times as expensive as baseband coaxial wire. However, there are a number of noteworthy benefits to using fibre optics. To begin with, it remains completely unaffected by any kind of electrical interference. Additionally, it is compact and easy to carry. Last but not least, it has a maximum distance of 30,000 feet and can support transmission speeds of 800 megabaud. Due to these characteristics, fibre optics are expected to see more industrial usage as technology advances (figure 1.9).

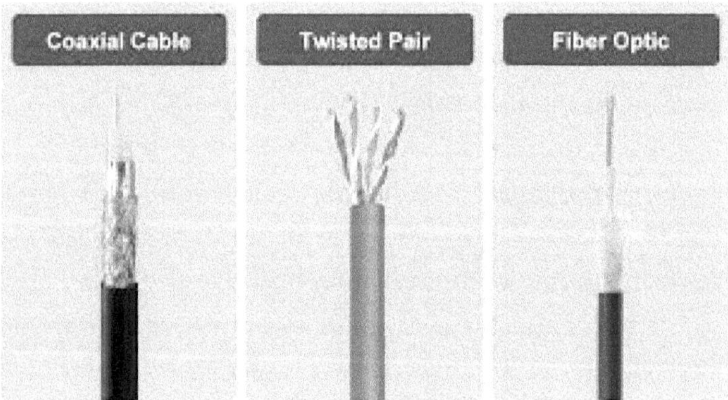

Figure 1.9: Three Types of Cabling
Source: - (Vcelink, 2024)

The role of software-defined networking (SDN) and network function virtualization (NFV).

Software-Defined Networking (SDN) and Network Function Virtualization (NFV) have transformed modern networking by shifting from hardware-centric architectures to software-driven solutions. SDN decouples the control plane from the data plane, which simplifies operations and allows for real-time adaptation; it also centralises network management, so administrators can administer the whole network via a single interface. NFV complements this by virtualizing traditional hardware-dependent functions like firewalls and routers, running them on general-purpose servers to reduce costs and expedite service deployment. Together, these technologies enhance scalability, flexibility, and efficiency, meeting the demands of advanced technologies such as IoT, 5G, and edge computing while addressing challenges like resource optimization and service delivery. This combination marks a significant evolution toward agile, future-ready networking ecosystems.

- **Software-Defined Networking (SDN):**

  Figure 1.10 shows that the control and data planes of conventional IP networks are closely connected, meaning that the networking devices used for both the data plane and the control plane execute their respective tasks. It seemed to be the best approach to provide network resilience, which was a critical design aim, and was therefore regarded significant for the early days of the Internet's creation. One major issue with this connected paradigm is its architecture, which is both complicated and rather rigid. Another concern is with the administration of the network, which is usually done by a plethora of proprietary solutions that use their own unique operating systems, control applications, and specialised hardware. Due to the high OPEX/CAPEX caused by operators having to buy and maintain several management systems and the specialised teams that go along with them, as well as the lengthy return on investment cycles and innovation limitations, this situation is not ideal.

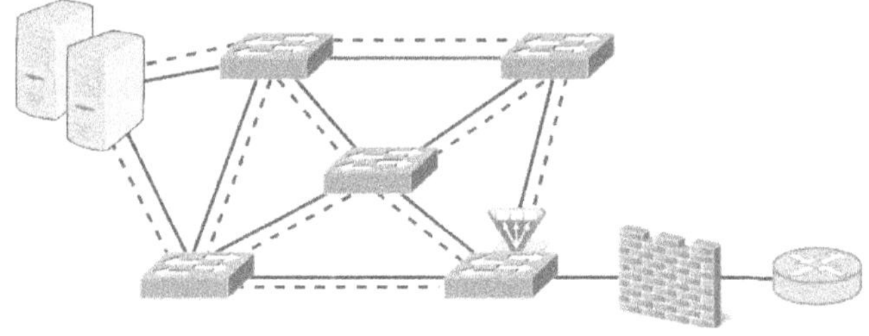

Figure 1.10: Conventional networking with coupled control\data planes

Source: - (Massimo et al., 2017)

- **Network Function Virtualization (NFV):**

The current state of hardware appliances is highly proprietary, and the expense of providing space and energy for several middle boxes makes it difficult to introduce new services into existing networks quickly. The concept of Network Function Virtualisation (NFV) represents a sea change for network operators when it comes to infrastructure design and deployment. NFV focusses on decoupling software instances from hardware platforms. As seen in Figure 1.11, the fundamental principle of virtualisation is the use of software virtualisation methods to create virtualised network functions (VNFs), which are then executed on commodity hardware, such as industry standard servers, storage, and switches.

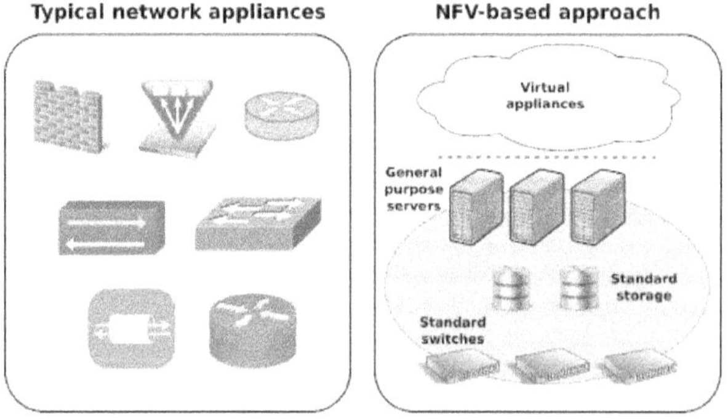

Figure 1.11: The network virtualization paradigm

Source: - (Massimo et al., 2017)

Telco operators stand to gain a great deal from the virtualisation concept. These include, but are not limited to: (i) reduced capital investment; (ii) energy savings through consolidation of networking appliances; (iii) faster time to market for new services through software-based service deployment; and (iv) introduction of customer-tailored services. Even more so, virtualisation and softwarization are two sides of the same coin that work well together. To ease compatibility with previous installations and improve performance, SDN may offer network virtualisation, for instance.

1.3 The Intersection of AI, Security, and Networking

The goal of AI networking security is to fortify networks against cyberattacks of all kinds. Viruses and malware are examples of more elementary dangers, whereas advanced persistent threats and complicated hacking efforts are examples of more advanced problems. Network security systems driven by AI can keep an eye on things like user behaviour, application use, and network traffic in order to spot any suspicious trends or abnormalities. Upon danger detection, the system has the option to notify administrators or initiate automatic responses to eliminate the threat. Repairing corrupted data, separating impacted systems, and banning malicious IP addresses are all part of this process. Another use of AI in network security is predictive analysis, which may help find weak spots and foretell future assaults by analysing trends in past data. In addition, these AI systems have the potential to improve their network security measures as they gain experience and knowledge.

AI in Networking: AI in networking or AI Networking represents a next generation of cloud-managed network management systems ('NMS'), incorporating an expanded set of advanced AIOps technologies with a refreshed NMS UX to optimize and automate the performance, security and management of network infrastructure. AI in networking also leverages advanced data science techniques such as GenAI transformers and DL classification AI on large network telemetry data sets to advance the efficacy of AIOps capabilities within networking. AIOps is modernizing the management and observability of critical IT infrastructure such as networking by leveraging intelligent automation and real-time analytics. When combined with a refreshed NMS interface designed for broader contextualized network experience observability, organizations can gain actionable insights that streamline operations and deliver superior user experiences, ensuring their networks are not just responsive but also resilient. Additionally, AI can enable proactive capacity planning by analysing usage trends and predicting future demands, allowing businesses to allocate resources effectively and maintain optimal performance as they grow.

AI in Security: Security technologies can employ AI algorithms for a variety of purposes, including pattern recognition (e.g., acoustic signals from gunshots), behaviour analysis (e.g., in surveillance systems), image classification and matching (e.g., using computer vision to distinguish between humans and animals), and material detection and identification (e.g., X-ray scanners for drugs or illegal substances). Increasing the likelihood and speed of detection, decreasing operator effort and weariness, and helping to concentrate the attention of security professionals where it is most required are just a few of the ways in which operational security may benefit greatly from the application of AI in security technology. AI has the potential to alleviate expenses, guide resource allocation, bolster decision-making, and even provide early intervention chances to counter insider threats at the managerial level. The usage of even the most basic forms of AI is integral to many security systems. Technologies that use AI to do many tasks at once may give the impression of being smarter than they really are. The amount of tasks that a system or device can execute is not always a good indicator of its degree of AI; rather, intelligence is often enhanced by the integration and complexity of decision making.

1.3.1 Why Integration is Critical

- **Evolving Threat Landscape:** The proliferation of sophisticated cyber threats has rendered traditional security measures insufficient. Malware, ransomware, and advanced persistent threats (APTs) are now evolving at an unprecedented pace, often leveraging AI themselves to bypass conventional defences. Networking environments, too, have grown in complexity with the advent of cloud computing, the IoT, and 5G technologies. These advancements have expanded attack surfaces, making security a more dynamic challenge. AI integration offers adaptive and intelligent solutions capable of detecting anomalies, predicting threats, and mitigating risks in real time.

- **Real-Time Decision Making:** Networking and security require real-time decision-making to ensure seamless operations and protect against breaches. AI-powered tools excel in processing massive volumes of data at speeds unattainable by human analysts. For example, AI systems may examine data transfer patterns across networks, spot irregularities that might indicate security breaches, and set off automatic countermeasures. This rapid action not only minimizes downtime but also reduces the impact of attacks.

- **Enhanced Visibility and Control:** AI-driven analytics provide unparalleled visibility into network and security environments. By correlating data from diverse sources, AI enables organizations to gain deeper insights into system vulnerabilities, user behaviours, and potential entry points for attackers. This level of transparency is crucial for designing robust security policies and optimizing network performance.

- **Scalability and Efficiency:** As networks scale to accommodate growing user demands, manual management becomes impractical. AI integration facilitates automation of routine tasks such as device configuration, network monitoring, and threat response. Both efficiency and the allocation of human resources to more strategic endeavours are boosted by this automation. Furthermore, AI's ability to learn and adapt ensures that security and networking solutions remain effective in dynamic environments.

- **Breaking Down Silos:** In the past, the concepts of security and networking have been isomorphic where security and networking are for different domains: while security is mostly implemented and administrated by a dedicated team and has separate branches, computer networking has its nodes and links as well. Such an approach is quite unconstructive and isolating, thus creating those infamous blind spots in the defences. AI is unifying in its essence and effectively ties these domains together by sharing a common space for threat intelligence and operationalization. This kind of approach is especially needed to protect against multi-vector threats that affect the network as well as applications and data.

- **Future-Proofing Technology Ecosystems:** Technological advancement is fast and thus requires better solutions to be adopted in the market. Because AI can make predictions, it allows an organization to expect what else could happen to networking and security in the future. The ML models can be used to replicate attacks, experiment with anti-attack measures, and enhance protection approaches all the time. This way, the proactive approach is guaranteed to provide immunity to other threats as they emerge into the market.

1.4 Real-World Applications of Integrated Systems

AI, security, and networking integration is ongoing in different industries and sectors and it demonstrates what is possible when changing the traditional business and organization models. Below are key real-world applications, supported by academic research and industry practices:

1. Smart Cities:

The goal of the Smart City is to use the technological advancements of the last several decades to improve the economic standing and way of life of city dwellers. Various smart city frameworks are presently accessible as a consequence of efforts made by the government, specialists, and academics worldwide. All possible aspects of city operation are covered by the various components of the framework. In this paragraph, we will examine the function of AI in identifying the crucial components from various frameworks. Artificial Intelligent (AI) and network systems are critical enablers to smart cities. These systems increase public safety with AI based tools for surveillance that incorporates real time anomaly detection and predictive policing. For instance, smart city solution where IoT devices and intelligent algorithms analyse traffic details, power use and control civil infrastructure to promote optimum usage.

2. Healthcare Systems:

AI has initiated a transformational age in healthcare, facilitating advancements in precision medicine, predictive diagnoses, and tailored therapy. AI provides optimal information for clinical decision-making, aiding in the prediction of illness development, identification of viable therapy routes, and enhancement of patient care. AI technologies, including ML techniques, are used to facilitate intricate healthcare functions such as diagnostics, patient management, and administrative procedures. AI has initiated a paradigm shift from a uniform approach to a personalized healthcare model by using extensive databases, resulting in therapies tailored to specific patient characteristics. This trend has been essential in establishing a future of precision medicine, where results may be enhanced by addressing the individual requirements and circumstances of patients (figure 1.12).

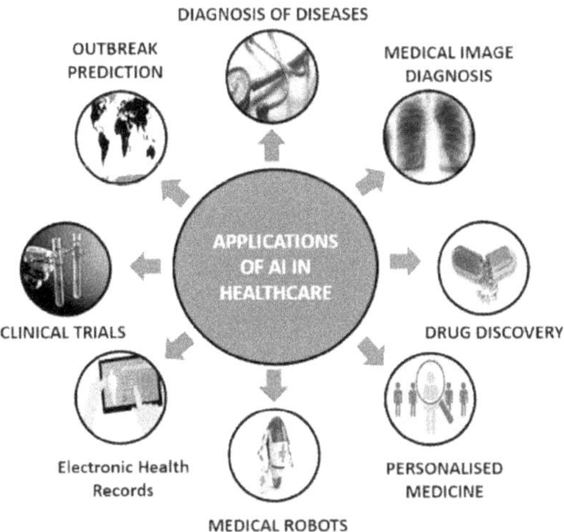

Figure 1.12: Application of AI in Healthcare

Source: - (Pandya, 2021)

3. Financial Services:

In financial institutions, AI-integrated systems bolster fraud detection and secure networking. ML models analyse transaction patterns to identify anomalies, while secure networks prevent unauthorized access to sensitive information. The adoption of these systems has reduced fraud by up to 30% in some banks. A new capacity is reshaping the financial sector: the ability to gather massive amounts of data from the environment and interpret it using AI and ML. AI and ML allow for better economic, financial, and risk event prediction, market transformation, risk mitigation, compliance, and prudential supervision. These advancements also provide central banks with more resources to carry out their monetary and macroprudential responsibilities. This technology has the potential to improve several areas of operations for financial services businesses, such as data analytics, forecasting, investment management, customer service, risk management, and fraud detection. AI is modernising the financial sector by increasing our understanding of financial markets, automating banks' once laborious processes, and creating customer engagement methods that mimic human intelligence and interaction. As it revolutionises the way banks operate, AI is also propelling new businesses. Faster and more accurate transactions, richer insights, and investment destination dictation are all outcomes of AI models using real-time market data. Security, fraud, AML, KYC, and compliance programs are all aspects of risk management that financial organisations may enhance with the use of AI technologies (figure 1.13).

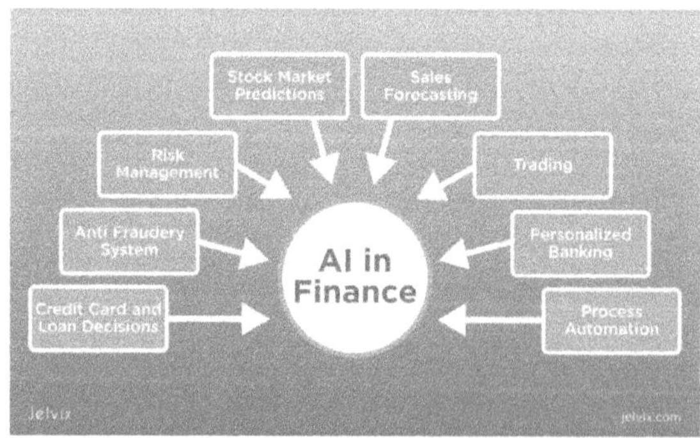

Figure 1.13: Uses of AI in Finance

Source: - (Fry, 2023)

4. Industrial IoT and Manufacturing:

Industry 4.0, marked by the emergence of neural networks (NN), is the fourth industrial revolution (Figure 1.14). Integrated AI and security systems have transformed manufacturing by ensuring secure communication between industrial IoT devices and central systems. Predictive maintenance powered by AI prevents downtime, while secure networking minimizes risks of cyber-physical attacks. The IoT is extensively utilized in smart manufacturing processes, with numerous applications contributing to improved productivity, streamlined operations, and safer environments. This piece discusses numerous smart factory IoT applications. Intelligent factories use IoT for machine-to-machine connectivity and real-time data. Impactful testing, analysis, prediction, and recommendation technologies enable intelligent production using these insights. Sensors in industrial facilities collect machine data, which centralised data platforms analyse and recommend. Recent years have seen AI and IoT gain prominence in numerous applications. Sensors, smart devices, cloud storage, and the internet exchange real-time data in IoT. AI uses ML to analyse real-time data and enable intelligent behaviour. In agriculture, healthcare, transportation, stock markets, and smart cities, AI and IoT are used. Industry 4.0's smart manufacturing can alter the industry with AI and IoT. Smart factories are efficient and high-quality by combining industrial processes with communication and information technology. NNs automatically identify, classify, and forecast results from enormous amounts of organised or unstructured data. IoT, big data (BD), and AI are revolutionising our everyday lives with technologies like Google Home, autonomous cars (Tesla), and drone deliveries.

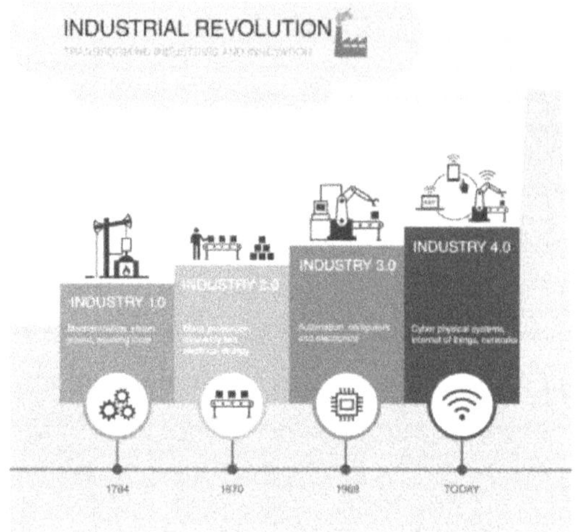

Figure 1.14: Four industrial revolution

Source: - (Mori, 2022)

5. Defence and National Security:

The use of AI in defence and security is a watershed moment in the evolution of global strategic settings. AI has emerged from the realm of science fiction and is now firmly planted in the real world. It is poised to impact security measures, military strategies, and the global power dynamic. Military and defines sectors utilize AI-powered networking systems for secure communication, threat detection, and autonomous weapon systems. AI ensures real-time monitoring of sensitive areas, while secure networks facilitate command and control in mission-critical scenarios. At a period of exceptional technological advancement, the intersection of AI with security and defence is a ground-breaking frontier that will transform military strategies, security laws, and the global power dynamic.

6. Retail and E-Commerce:

AI is rapidly transforming the retail industry, reshaping how businesses operate and interact with customers. Specifically, experts predict that AI in retail would reach $85.07 billion by 2032, up from $9.36 billion in 2024, a CAGR of 31.8%. As consumers increasingly demand personalized shopping experiences, retailers are turning to AI to meet these expectations. From dynamic pricing to automated customer support, AI technologies are enhancing efficiency and improving customer satisfaction. Retailers leverage integrated systems for personalized customer experiences and supply chain optimization. AI-driven insights enhance customer recommendations, while secure networking ensures the safety of payment gateways and customer data.

Multiple Choice Questions (MCQs)

1. **What is the primary goal of artificial intelligence (AI)?**

    a. To mimic human behaviour and intelligence
    b. To automate repetitive tasks
    c. To replace all human roles
    d. To develop synthetic biology

2. **Which of the following introduced the term "Artificial Intelligence"?**

    a. Alan Turing
    b. Marvin Minsky
    c. John McCarthy
    d. Herbert Simon

3. **Which AI subfield focuses on categorizing objects based on data features?**

    a. Natural Language Processing (NLP)
    b. Machine Learning (ML)
    c. Deep Learning (DL)
    d. Generative AI (GAI)

4. **Which type of natural language processing application converts spoken language into text?**

    a. Text-to-Speech (TTS)
    b. Speech-to-Text (STT)
    c. Speech-to-Speech (STS)
    d. Text-to-Text (TTT)

5. **What does GPT in GPT models stand for?**

    a. Generative Pre-trained Transformer
    b. General Processing Transformer
    c. Graphical Pre-trained Text
    d. Generic Predictive Training

6. **Which of the following is a key function of a network?**

    a. To store information
    b. To share information between members
    c. To increase computing speed
    d. To monitor hardware performance

7. **What is the primary role of Ethernet in networking?**

    a. To provide wireless connections
    b. To enable local area network (LAN) connections
    c. To connect long-distance networks
    d. To enable cellular data transmission

8. **Which Wi-Fi standard has the highest transmission speed?**

    a. IEEE 802.11a
    b. IEEE 802.11b
    c. IEEE 802.11g
    d. IEEE 802.11n

9. **What is the primary advantage of IEEE 802.11a over IEEE 802.11b?**

    a. Higher maximum speed
    b. Better interference resistance
    c. Longer range
    d. Compatibility with older devices

10. **In which network type does data transmission occur over long distances, typically via public telephone networks or satellite connections?**

    a. LAN
    b. WAN
    c. MAN
    d. WLAN

Answer

1	2	3	4	5	6	7	8	9	10
a	c	c	b	a	b	b	d	b	b

Chapter 02

INTELLIGENCE IN NETWORKING SYSTEMS

2.1 Overview of Artificial Intelligence's Role in Modern Networks

The advancing and intricate nature of current communication systems requires a new approach towards the management and tuning of such networks. In today's globalized world with an increasing number of Internet of Things (IoT) devices and continuous expansion of 5G networks, the expectations from current and future networks are significantly higher. The increased data volume, combined with the heterogeneity of users and applications, has exposed the drawbacks of conventional network control methodologies. These methods, which hinge upon the definition of static policies and manual settings, often do not meet the evolving complexity of modern networks. The applicability of AI-based models for enhancing the scalability of a network has been investigated. In this paper, the author discussing how organizational networks expand in size and complexity, more conventional approaches to network management fail. Instead, Al uses machine learning to automate repetitive tasks and to generate complex reports for business decision making. For instance, real time learning has been used to design adaptive routing that can complement the conventional systems by adapting to the actual flow of data on the network floor in order to increase efficiency and reliability of the system. AI, the protagonist in the extensive process of reinventing the means of communication networks. Al, with rich machine learning, deep learning, and neural networks, provides an approach to solve the numerous problems encountered in these networks. Whether it be in improving speed and productivity or building up security and functionality, AI holds the utilities that support and advance the handling of communication systems. This is not an enhancement but a transformation towards intelligent, adaptive and self-managing networks applying AI in this domain. AI-powered network automation tools and platforms represent a significant advancement in managing and optimizing communication networks. By automating complex tasks, providing predictive insights, and enhancing overall efficiency, these tools are reshaping the landscape of network management. As technology continues to evolve, AI-driven automation will play an increasingly central role in ensuring that networks remain resilient, efficient, and adaptable to future challenges (Esenogho et al., 2022).

The Need for Intelligence in Networking: Challenges and Opportunities

Anomaly detection and pattern recognition share certain similarities, but they are distinct processes with distinct ends in mind. In pattern recognition, the goal is to unearth hidden features in data, whether they are explicit or latent. Algorithms may be trained to recognize different types of data that have these traits by distilling them into feature sets. In a similar vein, anomaly detection takes a counter-approach to knowledge discovery. Establishing a sense of normalcy that describes most (say, more than 95%) of a given dataset is more important than discovering specific patterns that exist within certain subsets of the data. Any subsequent change from this normalcy will be considered an anomaly. The process of identifying and distinguishing between normal and aberrant patterns is sometimes misunderstood as anomaly detection. The data utilised to train the pattern recognition algorithm must be the only source from which patterns may be extracted. In contrast, anomaly detection allows for the possibility of an endless number of outlier patterns, including those originating from hypothetical data that is not actually present in either the training or testing datasets. Perhaps the most well-known application of pattern recognition is spam detection. This is due to the fact that spam usually exhibits a predictable set of characteristics, which can be used to train an ad algor algorithm to classify emails based on this pattern. On the other hand, spam detection can also be seen as an anomaly detection problem. In theory, we have succeeded if we can derive a set of features that adequately describes normal traffic and labels massive outliers as spam. In practice, though, spam detection may not be a good fit for the anomaly detection paradigm; after all, it's easy to see how spam messages often share more similarities than the vast majority of normal traffic. Additional applications that clearly fall under pattern recognition include malware detection and botnet detection. ML is particularly helpful in cases where attackers use polymorphism to evade detection. Fuzzing refers to the act of introducing arbitrary inputs into software in order to cause it to crash or enter a vulnerable mode that can be exploited later on. Inexperienced fuzzing efforts frequently encounter the challenge of iterating across an unmanageably huge application state space. Fuzzing is now significantly more efficient than blind iteration, thanks to optimizations in the most popular fuzzing software. Such optimizations have also made use of machine learning, which has learnt patterns of vulnerabilities in related programs and directed the buzzer to code routes or idioms that are similarly vulnerable, allowing for possibly faster results (Chio & Freeman, 2018).

Challenges of AI in Networking

Even though monitoring AI networks in a production setting might be beneficial, there are several obstacles that could prevent a successful deployment. In order for IT operations teams to properly deploy AI monitoring, they must first overcome a learning curve. The following issues should also be taken into account by network operations teams when using AI network monitoring:

- **Data quality.** AI can only effectively distinguish between innocuous abnormalities and real network issues with the data it is fed. AI network monitoring could be erroneous if sufficient real-time data is not collected or if data integrity is compromised. Artificial data, made to mimic a production network, is a popular substitute for real-time data streams in many organisations. While synthetic data does resemble real network activity, it is not the same and could either overlook a genuine problem or identify an issue that does not actually exist.

- **Integration.** It might be challenging to integrate new technologies into an existing production environment›s monitoring and management systems. The results can be lacking if the IT department is unable to integrate its AI technologies for monitoring networks with other management and security services.

- **Ethical issues.** Some worrying ethical concerns around abuse and operational integrity may arise as AI tools become smarter. This is also true in network monitoring, another area where AI is analyzing massive volumes of data. Regulating AI's data usage is a challenging and ever-changing endeavor, but governments are attempting it.

Benefits of AI Network Monitoring

Autonomous vehicles and smart speakers that process natural language are two examples of non-IT management use cases where artificial intelligence has advanced as an essential component. These days, more and more people are looking to AI to help with IT operations. AI network monitoring has various possible benefits that lead to a more optimised, dependable, and productive business environment. The following are some of the advantages of AI network monitoring:

- **Improved and simplified IT monitoring.** The density of linked equipment and components, as well as the virtualised character of networks, are still evolving. The task of monitoring these ecosystems has grown complex and frequently ineffective. It can be very difficult to find problems before they have a detrimental impact on operations. IT frequently searches through several monitoring technologies to identify the causes of service-level degradation. However, by offering a clear view of every component of the infrastructure as it functions together, AI network monitoring promises to dispel the fog surrounding network activity tracking and streamline IT processes. Quickly identifying and fixing the source of an issue is possible with the help of AI monitoring, which can also improve root cause analysis.

- **Security insights.** Rapid and accurate threat identification and speedier remediation are two other benefits of AI monitoring that can help throw light on any security vulnerabilities. This not only helps businesses strengthen their security measures, but it also decreases the number of false positive signals that security analysts receive.

- **Driving automation.** AI can provide guidance to systems that can automate formerly manual tasks. Some companies are looking into automating higher-level procedures, while others may be iterative continuous maintenance support components. Automated mitigation and remediation solutions are one application of AI (Kaur & Khan, 2022).

2.2 AI-Driven Network Optimization

AI techniques in wireless network security are at the forefront of safeguarding these dynamic and complex environments. These techniques harness. the capabilities of AI, including ML, DL, NLP, behavioral analysis, predictive analytics, anomaly detection, and more, to address the evolving threat landscape. Machine learning, for instance, enables the identification of known attack patterns, offering accurate threat classification and early warnings. Machine learning specifically deep learning does an excellent job in identifying complex patterns in network traffic and in real-time detection of subtle new threats. For text data classification, therefore the categorization of security incidents and the handling of these incidents, NLP is useful. Behavioral analysis sets the normal behavior of devices and users so any change, which may be indicative of fraud, is easily identified. In contrast, predictive analytics focuses on trends and happenings of the past with a view to anticipating perils. In real-time, anomaly detection systems are always monitoring the network traffic and the users and labeling anything that is different from the norm as a threat. The value of these techniques lies in their adaptability and responsiveness, Artificial Intelligence for Wireless Communication Systems allowing for dynamic defence strategies in a dynamic network environment. These AI techniques improve the security of next generation wireless networks by analyzing large volumes of data to learn patterns and identify anomalies thus enabling threats to be identified early and security mechanisms to evolve with the progressive threat landscape of wireless networks.

The integration of AI into wireless network security extends beyond. threat detection and privacy preservation. Due to the Self-organizing characteristic of AI, feasible network management methods can improve resource allocation, QoS, and network performance. These techniques employ AI to speculate patterns in traffic, identify areas prone to congestion and self-organize resources to enhance a seamless network operation. Furthermore, AI-implemented and reinforced network management is capable of detecting signs of security violations or system degradation. When used together, security and network management can provide a centralized point of control over operators' networks as well as protection against threats (Sur et al., 2024).

2.2.1 Reducing Latency with AI

Identifying Latency Bottlenecks with AI Models

The term 'Latency' with regards to AL models and systems is the time between receiving an input and producing the resultant output. This measure is most relevant in the contexts where near real-time performance is critical, for instance; self-driving cars, smart live-chat systems, or trading bots. Latency is normally expressed in milliseconds (MS) and may be controlled by using profiling tools that record the amount of time each activity in the models takes, with the aid of specialized techniques such as parallelism. Both load balancing algorithms and resource management are important for multitasking performance. In practice, the benchmark for effective multitasking varies depending on the specific application, the tasks' complexity, and the design of the AI system. There isn't a one-size-fits-all percentage; each AI application may set its own performance targets based on its unique requirements and constraints. Success in multitasking enhances AI systems' versatility and real-world applicability, making them more efficient and adaptable to complex environments (Smith et al., 2024).

The architecture of Spark features a distributed computing framework that will help reduce latency and optimize resource allocation for easier management and building of scalable AI applications by handling large-sized data sets.

Ray: Ray is a distributed computing system designed to scale AI and machine learning workloads. Ray efficiently enables one to perform job scheduling across multiple nodes, carry out the distributed training of deep learning models, and maintain a simple and flexible API for developing distributed applications. Further, it integrates with popular deep learning and machines.

Adaptive Routing Techniques to Minimize Delays

Optimisation of data transmission, resource utilisation, and overall network efficiency are greatly aided by adaptive routing and load balancing, two fundamental ideas in the field of communication networks. The process by which data packets are chosen to travel across a network on an as-needed basis is called adaptive routing. Adaptive routing procedures alter at any given time and occur through the present state of the network as compared to the static routing approach, which employs set paths. For that reason, the system can adapt to the new conditions, for example, network congestion, breakdowns or changes in traffic rate. Adaptive routing algorithms work with the goal of increasing the network's reliability, maximizing throughput, while at the same time minimizing delays through constant monitoring and examination. However, load balancing refers to the technique of distributing traffic on a network over used resources or channels to prevent overwhelming. Load balancing is key essential in communication network scenarios since it enhance system efficiency and prevent delays. Load balancing algorithms are used in optimizing the utilisation of some resources and curtailing response time in handling data packets by directing them through several network channels or servers. Combining adaptive routing with load balancing shows how the two work together to influence network dynamics. When applied with load balancing methods, the adaptive routing algorithms keep on analyzing the overall health of the network and redirect traffic in order to optimize throughput while keeping latency as low as possible. For instance, Adaptive Routing could take the traffic on the path to a less crowded path when a given path is congested. To add more reliability and efficiency to the communication infrastructure, the methods of load sharing are employed to avoid congestion in some specific constituent or path. All these concepts become even more important in the modern communication networks where data volumes and traffic patterns are growing rapidly. Not only does adaptive routing and load balancing enhance the quality of service for end users but it also enhances the scalability and reliability of the communication structures allowing them to meet the challenging demands of the digital communication outlets (Hammouch & Jamil, 2024)

Case Studies: AI-Driven Latency Optimization in Real-World Networks

For applications that experience burst traffic, latency is the key performance indicator that matters the most. When thinking about interactive and real-time services, latency is crucial. Control plane latency and user plane latency are the two most common ways to describe latency. The delay in the control plane involves the process of attaching to the network, whereas the delay in the user plane solely takes into account the delay of packets when the UE is connected. Online games and VoIP rely on low user plane latency. There is a display of the latency-related key performance indicators in Table 2.1. Ping performance will be utilised as a starting point for analyzing and optimizing the LTE network's latency profile. The user plane latency is being discovered via Ping. The delay is mostly affected by the following: the UE state, namely the RRC state, the payload, specifically the ping size, and whether it is an MTC or MOC. Figure 2.1 displays the LTE latency overview. On average, the time it takes for a data packet to go from the user plane to the physical layer is known as user plane latency. User plane delay in one direction is half of user plane latency in both directions. While the exact figure depends on factors like system loads and radio propagation conditions, LTE requires a minimum of 5 ms for one-way delay across the radio access network under ideal circumstances. This number is associated with an RTT of 10 MS, which is difficult to accomplish in reality. Transition times between LTE modes are measured by control plane latency, also known as connection setup delay. The foundation of LTE lies on two primary states: RRC connected (active) and RRC idle (non-active).

Table 2.1: The latency-related KPI

Control Plane		User Plane	
KPI Name	Critical	KPI Name	Critical
1st Page response time	Yes	Dedicated bearer activation time	No
UE activation time (idle to active)	Yes	VoIP call end time (Mobile to Mobile/ PSTN)	No
Service request set up time	Yes	VoIP Mobile-to-Mobile Mouth-to-Ear delay	Yes
MO VoIP call set up time (Mobile<>PSTN)	Yes	VoIP Mobile-to-Mobile Packet Latency	Yes
Mobile-to-Mobile VoIP call set up time	Yes	VoIP Mobile<>Land Mouth-to-Ear Delay	No
Ping RTT between UE and application server (unscheduled and prescheduled)	Yes	VOIP Mobile<>Land over Internet Packet Latency	No
Initial attach/detach time	No	VOIP Mobile<>L and over PSTN Packet Latency	No

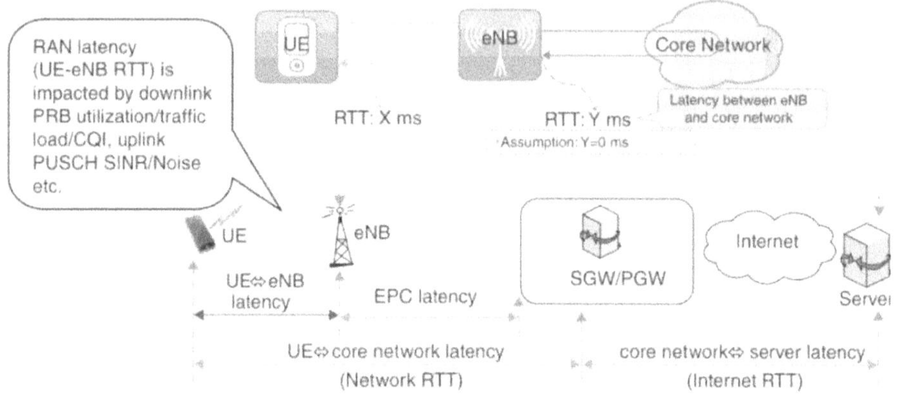

Figure 2.1: LTE latency

Source: - (Zhang, 2017)

2.2.2 Resource Allocation and Load Balancing

Al algorithms dynamically allocate network resources, like bandwidth and power, based on real-time traffic demands, improving network efficiency and performance. Al techniques balance network traffic across available resources, preventing congestion and optimizing throughput (Pawar & Ingole, 2024).

Load Balancing: To guarantee that no one resource is overloaded, load balancing divides incoming traffic and workloads among several resources. Performance and availability are improved. Load balancing and resource management are critical components in the deployment and operation of high-performance, scalable systems. As computational demands grow, efficient distribution of workloads and effective utilization of resources ensure optimal system performance, reliability, and user satisfaction. This chapter delves into the principles, strategies, and technologies involved in load balancing and resource management. highlighting their importance in modern computing environments. The practice of spreading workloads among several computer resources, like servers, in order to prevent any one resource from becoming overloaded is known as load balancing. This technique enhances the reliability and availability of applications by spreading the load evenly, preventing bottlenecks, and improving response times.(figure 2.2) (Neeraja et al., 2024)

- **Types of Load Balancers:**
- **Hardware Load Balancers:** Physical devices that manage traffic.

    - **Software Load Balancers:** Applications that perform load balancing functions.
    - **Cloud-based Load Balancers:** Cloud companies offer managed services to make deployment easier.

Load balancers can be configured to use various algorithms (e.g., round-robin, least connections) to efficiently distribute traffic.

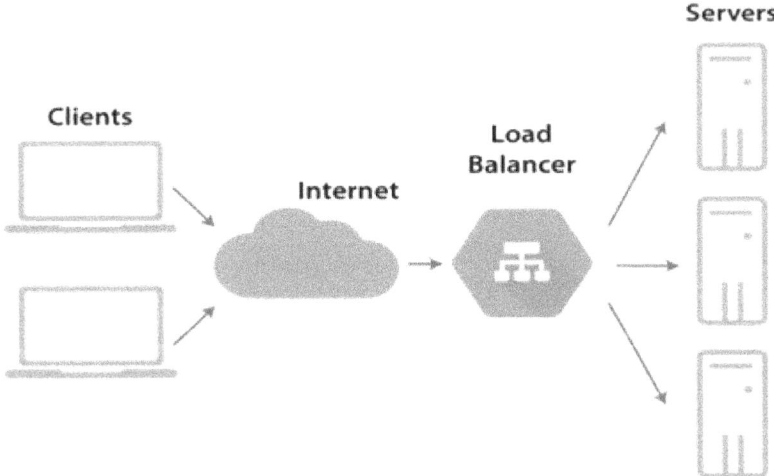

Figure 2.2: Diagram of load balancing
Source: - (Patil, 2023)

Resource Allocation: Efficient usage of resources within cloud environments for business requisites greatly determines the efficiency and reliability of a cloud infrastructure. Appropriate resource estimation assists organizations to optimally utilize organizational resources, satisfy user needs, and conform to SLAs. Fluctuations in loads and changing patterns of utilization require the efficient use of resources if the efficiency and quality of service delivery is to be sustained. In resource management, RI algorithms are employed to provide dynamic resources for compute, storage, and network for different components of a system, that includes microservices, virtual machines, and containers. So, the objects are to achieve the maximum resource utilization while satisfying quality of service (QoS) constraints including response time, latency and cost. The RI-based resource allocation strategies indicate that based on past usage patterns and other current trend information it is possible to forecast the requirement for the resources and thereby make timely allocation. For example, an agent established on DQN can identify the requirement for CPU and/or memory in various microservices in a cloud environment and in light of earlier and current workload of each microservice accordingly adjust their resource utilization and hence cut on cost while enhancing efficiency. In the same way, an agent under the PPO-based can be trained to manage bandwidth resources for various flows in the SDN to maximize networking efficiency and minimize network traffic. The goal of load balancing is to maximize efficiency and effectiveness by dividing up work among numerous computers, nodes, or containers using RL, algorithms. the figure 2 (Neeraja et al., 2024).

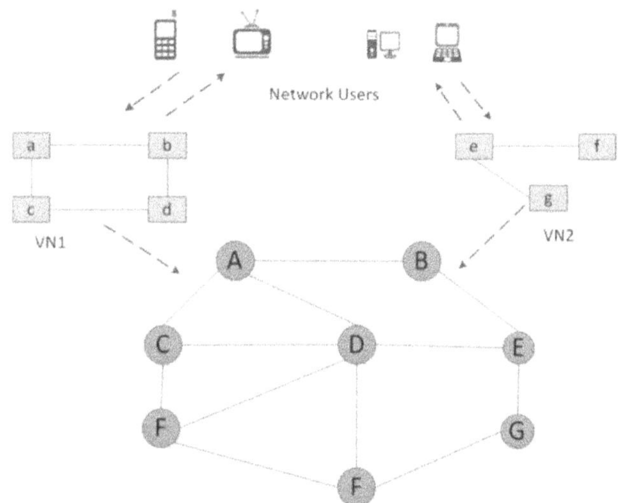

Figure 2.3: Resource allocation for networking

Source: - (Xiao et al., 2020)which has been proven to be an NP-hard problem.

At present, the methods to solve the problem of virtual network mapping still have the following defects. Most of the existing literature is limited to static virtual network (VN

Thus, load balancing is very crucial in DevOps architecture especially within cloud native and microservices based applications since the workloads are constantly fluctuating and unpredictable. Ri-based load balancing algorithms can be trained to shift the weight of load as per the monitor metrics like CPU and memory, response time, and network delay in real time. For instance, a Q learning-based agent can be trained in the case of processing HTTP requests in cloud environment to be distributed among multiple web servers in order to reduce the response time and improve the collection throughput. Similarly, a PPO-based agent can learn to balance workloads across nodes in consumption Kubernetes cluster, optimizing resource utilization and reducing energy (Tatineni, 2024).

First, resource allocation and load balancing are two interrelated techniques that ensure computational resources are shared equitably while their utilization is maximally effective. Dynamic resource allocation by tools like Kubernetes or Apache Mesos ensures that if there are changed workload demands, no one node will be a bottleneck. This is left for the algorithms of load balancing: not to overload and 405 X ensure good performance by splitting computing tasks and incoming traffic across a number of nodes.

Scaling techniques: This can be achieved through various scaling techniques, which involve volutionources manipulation so that they meet immediate needs. This is usually achieved in one of two ways: either by scaling out, which means adding more nodes, or through vertical scaling, where the nodes are made powerful. Seldom do cloud platforms like the Google Cloud AutoScaler and the AWS Auto Scaling dynamically scale up the resources where dynamic changes are in demand, variable loads for optimum performance with minimum human intervention.

Caching: Data are cached in the fast access storage layers to try to reduce retrieval times and hence improve performance. Techniques like data prefetching and in-memory caching may prove very effective in terms of latency reduction, thereby making the AI systems much more responsive, especially for high-throughput applications or real-time ones.

Algorithm and Model Optimization: In terms of performance improvement, Al algorithms and models should be optimized. That might include strategies to reduce computation and speed up the process of making inferences, pruning, deleting elements of the model not needed, quantization involving numerical precision reduction, and knowledge distillation-which may mean teaching smaller models to mimic bigger ones. Algorithmic modification and hyper-parameter optimization are other ways through which an improvement in model performance can be achieved.

Data Management: The effective and efficient data management results in better performance of Al that includes a number of tasks such as compression, indexing, and partitioning of data. On one hand, enormous amounts of data can be handled and computed in minimum time with the support of some data managing and optimization tool like Amazon Redshift and Apache HBase.

Network Optimization: The network should be optimized to higher performances and data transfer rate at lower latency. Delays over the network may be reduced by strategies such as content delivery networks, efficient network protocols, and compression of data. Monitoring utilities may also help in locating and weeding out bottlenecks to ensure error-free communications between nodes.

Performance Profiling and Analysis: Performance profiling tools analyze behaviors of Al systems while helping identify areas for development. Using other tools such as NVIDIA Nsight Systems and Intel VTune Profiler, developers are allowed to optimize their code and increase runtime efficiency of both GPU and CPU with extended, detailed performance information. Feedback Loops and Continuous Improvement: Both techniques make the performance optimization of AI/ML systems a continuous process. These iterate over changes in system configuration, resource allocation, and algorithms with the help of automated monitoring and analyses to derive useful insights (Kanthavel & Dhaya, 2024).

2.3 Machine Learning Algorithms for Traffic Management

Today traffic conditions are managed through Machine learning applications which collect real time data from the sensors and the cameras also from connected vehicles to control the traffic situation and reduce the incidences of traffic congestions and accidents occurring on the roads. Smart transportation resulting from advanced technologies such as predictive modeling, reinforcement learning, and neural networks helps make intelligent decisions in traffic signal control, route planning, accident occurrence prediction.

1. Reinforcement Learning for Congestion Control:

The SL algorithms forecast future traffic patterns using labelled training data, but the USL algorithms can find structures and patterns in the network traffic data. Network traffic prediction can also make use of time series forecasting algorithms that are built to deal with temporal data. As a result of their training on labelled congestion states in historical data, SL algorithms are able to forecast congestion levels and implement solutions. In addition, pattern recognition, understanding the root causes of congestion, and developing measures to alleviate it can be accomplished with the application of USL approaches (such as clustering and dimensionality reduction techniques). It is also possible to use RL algorithms to make real-time adjustments to network characteristics (such as routing and bandwidth allotment) in response to feedback about congestion levels (Celik & Eltawil, 2024)a new horizon beckons with the advent of 6G, seeking a profound fusion with novel communication paradigms and emerging technological trends, bringing once-futuristic visions to life along with added technical intricacies. Although analytical models lay the foundations and offer systematic insights, we have recently witnessed a noticeable surge in research suggesting machine learning (ML.

2. Use Cases:

- **AI - Powered Traffic Shaping and Prioritization** - Traffic classification and traffic prioritization is a major application of AI in the management of traffic in a network. While these networks process different types of data such as the streaming of videos in real-time, IoT device communications, vital data must always preempt less urgent traffic. Algorithms can identify traffic in terms of its nature and urgency and provide improved routing determinations. With data prioritization and bandwidth control through scheduling, Al guarantees the availability of resources to such latency-critical applications as voice or video calls and distributes the network load efficiently for overall throughput. The second key application is predictive traffic management. AI will help foresee congestion patterns depending on usage data and network status and hence gear up network resources. This prediction assists with load distribution, and AI can decide how data is to be distributed across different paths or servers, thereby avoiding congestion. In cloud computing environments, this predictive capability enables dynamic scaling of resources, ensuring that the network can handle periods of high demand without compromising performance.

- **Network Traffic and Its Characteristics** - Data that is present in an operational network is called network traffic. At all times in a computer network, communication devices constantly poll service providers for access to resources, and service providers constantly reply. At the precise second that a request is made, a resource might not be accessible. Requests, services, release data, and control data are forms of information that are constantly sent throughout the network. In this context, "data" refers to the vast quantities of information in the form of packets that are constantly circulating around the network. When the number of packets sent over a network exceeds the capacity of individual communication devices, congestion occurs. To maximize resource utilisation, a corporate network needs to regulate traffic within its boundaries. In order to manage network traffic, it is necessary to prioritize particular kinds of traffic, ensure a minimum bandwidth to other applications, and limit bandwidth to specific applications. "Traffic management" describes this practice. The typical application of traffic analysis is to define healthy traffic in order to determine the efficacy of resource utilisation. Based on the points of monitoring, there are primarily two kinds of traffic. They are known as LAN traffic and WAN traffic, respectively. LANs allow for the rapid and precise transport of data between devices within relatively small geographic regions. LANs and other small area networks are connected via WANs. In contrast to LANs, these LANs cover vast geographic areas at a slower data transfer rate. The utilisation of network resources and services causes WAN traffic to fluctuate more than LAN traffic. WAN data may be categorised as either bursty, random, or Internet-based. Intervals of seemingly random traffic seem to occur at regular intervals. The Poisson model describes the distribution of online traffic. In conclusion, bursty traffic is characterised by intermittent, abrupt spikes in activity. Network traffic characteristics are crucial for detecting or identifying malicious traffic. The packet level and the NetFlow level of network traffic analysis are both theoretically feasible. For the purpose of collecting and monitoring traffic data produced by routers or switches that are NetFlow-enabled, Cisco[2] launched NetFlow, an IP (Internet Protocol) traffic flow technology. Analysis of traffic using NetFlow is quicker than analysis of traffic at the packet level. This is because NetFlow traffic does not typically include as many parameters. The following are some key characteristics used to describe network traffic: (a) the amount of time or session duration; (b) the number of packets or flow level parameters, which is often high; (c) the volume of traffic; (d) the noise level; and (e) the traffic rate.

- **Network Traffic Anomalies** - A traffic anomaly is a data point that deviates significantly from the typical pattern of transportation traffic. When anything happens on a network that isn't typical, it's called a traffic anomaly. A rising tide of incursions and attacks originating from networks causes this. A rise in cybercrime in recent years. There are numerous potential causes of network risks, such as (i) malicious actions that exploit regular network services, (ii) an overloaded network, (iii) faulty devices, and (iv) vulnerabilities in various network characteristics including protocols and ports.

Sophistication in Network Anomalies and Their Detection

The goal of anomaly detection is to identify data patterns in network traffic that do not follow the norm. The significance of anomaly detection stems from the fact that, across many different fields of application, data abnormalities can yield substantial and frequently crucial actionable insights. For instance, a computer network's unusual traffic pattern may indicate that a compromised computer is transmitting private information to an unapproved host. Various factors might lead to anomalies in a network. There are two main types of network anomalies, as mentioned in: (a) those that affect performance and (b) those that affect security. A number of phenomena can arise from abnormalities in network performance, such as broadcast storms (when many nodes in the network simultaneously transmit a large amount of data), transient congestions (which last only a short while), babbling nodes (which emit an increasing amount of data), paging, and file server failures. An attacker or attackers could be responsible for security-related network anomalies if they deliberately flood the network with excessive traffic in order to commandeer the bandwidth and prevent legitimate users from receiving service. Network anomalies pertaining to security are the exclusive focus of our book. At this time, the gold standard for intrusion detection methods is anomaly-based network intrusion detection. It is now one of the main areas of intrusion detection research and development. ANIDS capabilities are being added to a variety of systems, and other novel approaches are being investigated. The field is still in its infancy, though, and significant problems must be resolved before it is feasible to implement ANIDS systems widely. We can now connect remote parts of the world via the Internet to share a large amount of information thanks to advancements in networking technologies. However, as was already indicated, dangers from attackers, spammers, and criminal businesses are also increasing at an accelerated rate in tandem with this improvement. By evading a host's or network's security measures, an intrusion typically aims to jeopardise a system's availability, confidentiality, or integrity. To keep the infrastructure of big businesses safe, security professionals employ intrusion detection technology. An IDS is a combination of hardware and software that keeps tabs on all the activity happening in a computer system or network and looks for any indications of unwanted or malicious traffic. Anomaly detection systems and abuse detection systems are the two main types of IDS. Only known threats can be identified by misuse detection using signatures that have already been generated and saved. Therefore, it is crucial to update dynamic signatures, which is why IDS providers often publish new attack definitions. However, the continuously increasing number of vulnerabilities and exploits is too much for misuse-based systems to handle. However, systems that use anomaly-based detection are made to identify any departure from typical behaviour profiles. For the identification of unknown or unique attacks without any prior knowledge, they are more appropriate than abuse detection. Typically, however, these systems produce a lot of false alerts. Anomalies or intrusions in network traffic can be detected using four popular machine learning techniques: categories: (i) supervised, (ii) semi-supervised, (iii) unsupervised, and (iv) hybrid. The supervised method involves building a model for prediction using a training set that includes both typical and malicious data. In order to identify the class to which any unknown data object belongs, it is compared to the model. Only instances of the normal class are used in the training data for the semi-supervised method. There is a lack of attack class labelling for data instances. The model for the normal behaviour class is built using many different ways. Any outliers in the test data can be located using this approach. The unsupervised method has the most promise for a broad range of applications as the model doesn't need training data. The results might not be as helpful, though, because there isn't enough labelled data. Lastly, a hybrid strategy typically takes advantage of all of the aforementioned to achieve efficient and effective performance in identifying large-scale network anomalies. Unless otherwise stated, every method relies on the underlying premise that the test data contains many more normal occurrences than abnormal ones. These methods become flawed if this assumption turns out to be false. alarm rates that are too high. Anomaly detection in the first two scenarios calls for training on labelled examples. While it would be ideal to have a high number of instances of both normal and attack training, this may not always be possible. Furthermore, it is an exceedingly challenging undertaking to generate a set of genuine normal examples including all the variances. This book delves into the creation of top-notch methods for detecting anomalies in network data. These methods can identify both known and undiscovered assaults, and they outperform rival approaches in terms of detection rate and false-positive rate.

Network Traffic Anomaly Prevention

Anomaly prevention in high-speed networks is no easy feat. The goal of NIPS is to stop harmful traffic from ever reaching a network in the first place. Instead of sounding an alert after the fact, these technologies automatically discard harmful traffic before it can do any damage to the network. NIPS are effective and scalable in the same way as firewalls are because they verify the validity of incoming network traffic through deep packet inspection. Identical signature-based and anomaly-based IPSs are the two main categories of IPSs, just as IDSS are. To prevent known attacks, a signature-based NIPS use previously established signatures. The safeguard checks the signatures with an algorithm that matches patterns quickly. The goal of an anomaly-based NIPS, in contrast, is to thwart zero-day attacks, which involve hostile communication that is both novel and unknown. Anomalies in network traffic can be detected and prevented using clustering, classification, and outlier mining, three data mining jobs that are highly beneficial. Data mining, particularly in the area of network traffic analysis, has recently come to light as having the potential to greatly affect network security. It is crucial to accurately identify intriguing patterns from massive datasets and rapidly analyze them using excellent data mining techniques as the development of most security systems is dependent on these patterns. A well-executed data mining strategy can identify and thwart both known and unexpected threats (Bhuyan et al., 2017).

2.4 Predictive Analytics for Network Health and Performance

Metrics for the health of a network are numerical indicators of the network's dependability, efficiency, and performance. These metrics give information on the current status of the network and help organizations, to know the current state of their network. Since the delivery of applications and services in the context of the cloud heavily depends on the network's performance, monitoring these KPIs is vital to efficient service delivery. The definition of network health metrics can be broken down into several key components:

1. **Quantitative Measurements:** Network health metrics rely on numerical figures that enable simple comparison over period of time, where necessary. These measurements can include any number of parameters such as bandwidth, delays, frame, or bit errors, and data rate.

2. **Performance Indicators:** Evaluation indicators are used to measure the performance level of the entire network. For example, high latency may actually signify that there is some problem with the network while low errors mean it is a sound environment.

3. **Real-Time and Historical Data:** Network health metrics can be measured and captured one time or continuously. Real time measurement makes it possible to notice right away that there is something wrong or the performance is not as good as before while a historical measurement helps to notice trends and Store trends over time.

4. **Actionable Insights:** The general reason for monitoring networking health metrics is to get intelligence that can be used in problem solving. When the numbers are comprehended, the necessary changes can be made to improve efficiency, improve convenience, and manage resources.

<u>**Key Network Health Metrics**</u>

To gain a deeper understanding of network health metrics, it's essential to explore some of the most commonly used metrics in cloud environments:

1. **Latency:** The amount of time it takes for a data packet to go from its origin to its final destination is called latency. Most commonly, milliseconds (MS) are used for measurement. The application response time can be relatively high and slow due to high latency which is not good for users.

2. **Throughput:** The term "throughput" describes the rate at which data may be sent across a network; it is commonly expressed in bits per second (bps). If the network's throughput is high, it means it can efficiently process a lot of data.

3. **Packet Loss:** Uncompleted data transfer is experienced when data packets fail to get to their intended destination and is called packet loss. This statistic is very useful for analyzing network reliability because high value of packet loss can cause extra retransmission and, therefore, lower network performance.

4. **Bandwidth Utilization:** Currently, a specified percentage of the available bandwidth is being used, which is measured by bandwidth utilisation. It assists organizations in knowing whether they have made excess investments in the networks or whether the networks are underutilized.

5. **Error Rate:** The error rate is the number of packets that were transmitted in error on the data link layer. If the error rate is high then there must be some problem with a network device or the setup of a network.

6. **Jitter:** Jitter is a measure of the error that can occur when packets are late. For high-quality, real-time applications like audio and video conferencing, this is of the utmost importance.

7. **Connection Time:** Connection time denotes the time taken to set up a connection between two points. It is very important in evaluating the interactivity of applications and services. Through using these metrics and appreciating their importance, the network health condition can be regulated, and the right decisions concerning the cloud can be made.

Machine Learning and AI-Powered Monitoring The integration of ML and AI into network health monitoring has emerged as a game-changer for organizations. These technologies enable advanced data analysis, anomaly detection, and predictive analytics.

- **Anomaly Detection:** It also means that ML algorithms can look at historical network data to set up normal behavior as a reference point. Such Novel Approaches for Monitoring and Managing Network Health in Cloud baselines should always be compared with real-time data to detect discrepancies that can suggest emerging problems like security breakout or reduced performance.
- **Predictive Analytics:** AI-empowered monitoring enables work-up of network performance trends from past data for future prediction of issues that may hinder organizational operation hence solving them before emergence.

Modern NPM solutions offer comprehensive visibility into network performance by combining traditional metrics with advanced capabilities. These solutions provide real-time insights into network health, user experience, and application performance.

- **End-to-End Visibility:** Novel NPM tools can monitor performance across all layers of the cloud stack, from the infrastructure to the application layer. This end-to-end visibility helps organizations identify the root causes of performance issues more effectively.
- **User Experience Monitoring:** Advanced NPM solutions can assess the quality of the user experience by measuring metrics such as page load times and transaction success rates. This focus on user experience aligns network performance management with business objectives.

Techniques for Measuring Network Performance

Measuring network performance is a fundamental aspect of managing network health in cloud environments. Various techniques have been developed to capture and analyze key metrics that reflect the operational status of a network. These techniques enable organizations to identify performance issues, optimize resource allocation, and enhance overall user experience. This section explores several effective techniques for measuring network performance.

1. **Active Monitoring**: Performance monitoring entails the proactively sending packets through the network while periodically measuring parameters such as latency, Packet loss, and throughput, etc. This approach mimics normal usage of the network in order to determine the effectiveness of the network during the operational use.

   The benefit of active monitoring is that it can identify problems as they develop, and offers some perspective on how the network works at any given time. With artificial traffic, one is able to test specific paths in the network and diagnose conditions that are ailing it. Tools: Tools such as Ping. Active applications include; Traceroute and other network performance monitoring tools such as SolarWinds or PRTG. These tools can help get metrics like the round-trip time or even the hop count, information which is very useful for performance testing.

2. **Passive Monitoring:** Passive monitoring is different from active monitoring in that traffic is only examined as it passes through the network. It passes through and analyzes packets to obtain data on several performance parameters without creating other traffic. A primary advantage of passive monitoring is that an organization can obtain a complete picture of the network performance and watch for tendencies. Its performance is particularly high with all problems linked to packet loss, throughput, and usage of overall bandwidth. Tools for passive monitoring are Wireshark and NetFlow analyzers. With these tools, it is possible to gather broad data about the network traffic and, besides, get more specific data

on users and applications. Virtualization enables organizations to increase the capabilities of their resources at short notice. Since services demand varies over time, organizations can easily scale up or scale down of VMs to meet the workload demand. This elasticity is important in cloud systems where workload can fluctuate greatly.

Impact on Network Health:

The capacity to flexibly scale resources improves the health of the network by keeping applications responsive even during periods of high usage. Organizations can avoid performance degradation by scaling up resources as needed, providing a seamless user experience.

- **Improved Reliability and Fault Tolerance**

  Virtualization enhances the reliability of cloud environments through features such as live migration and automated failover. With live migration, VMs can be transferred between physical servers in real time, allowing organisations to respond to hardware breakdowns or execute maintenance without affecting service availability.

Impact on Network Health:

Improved reliability contributes to overall network health by minimizing downtime and ensuring that applications remain available to users. Automated failover mechanisms help maintain service continuity, further enhancing user satisfaction.

- **Simplified Management and Orchestration:**

  Virtualization simplifies the management of cloud resources by providing centralized control over VMs and infrastructure. Management tools enable organizations to monitor performance, allocate resources, and automate provisioning processes.

2.5 Autonomous Systems and Their Role in Networking

Big data must be considered throughout the design process of autonomous networks. Big data is defined by more than just data size; Doug Laney of Gartner has added to this definition by proposing the "3Vs" of volume, variety, and velocity. Variability, validity/veracity, visibility/visualization, and value are additional characteristics that have been added to the 3Vs to 5Vs or 7Vs by others (Livi13, Data14, Wikibig) since then. As will be explained in subsequent chapters, these characteristics are crucial for creating a high-precision AI system.

A variety of sources, including apps on the device itself, consumer edge devices like set-up boxes, network edge nodes like data centers, or even more centralised private and third-party clouds, can provide large data to autonomous networks. Keeping data of a good quality is difficult regardless of its source. The criteria of quality encompass things like accessibility, usefulness, dependability, appropriateness, and display. Problems with data management also exist. A large data system necessitates the determination of numerous data management and policy concerns, including the data types collected, retention durations, data summarization or anonymization methods, and batch processing versus real-time data access. It is costly to transfer data between different locations. Therefore, analytics and ML are typically applied to data by most modern systems, rather than the other way around. In order to facilitate parallel processing, data is usually preprocessed. One of the most prominent open source ecosystems for handling and consuming large data is Hadoop.

An autonomous network must be cared for and fed data despite the many difficulties associated with data handling, management, and processing. Cybersecurity, fault and performance management, data traffic flow management and optimisation, and the prediction of optimal placements for edge clouds, micro and macro cells all rely on data. Data is just as vital as money in today's digital society. Information must be handled with the same care and attention as money is. Today, most businesses implement data governance to establish guidelines for the best ways to handle, store, process, and safeguard data in order to lower operating costs, spur ML and AI innovation, and maintain compliance with data protection laws.

The foundation of ML models and the reason they are still in use today is data. Machine learners get more accurate the more varied the training data's properties are. On the other hand, inaccurate learners may result from poor and biased data. We call this algorithmic bias. When developing AI systems in SDN, high-quality, objective data is essential. This book has several chapters devoted to data integrity and quality (Gilbert, 2018).

Multiple Choice Questions (MCQs)

1. **Which of the following is NOT a key component of network health metrics?**

   a. Quantitative Measurements
   b. Performance Indicators
   c. Subjective Analysis
   d. Real-Time and Historical Data

2. **What does "latency" measure in the context of network health?**

   a. The percentage of data packets that fail to reach their destination
   b. The variability in packet arrival times
   c. The time taken for a data packet to travel from source to destination
   d. The total bandwidth utilized at a given time

3. **What is the primary role of machine learning in anomaly detection?**

   a. Generating synthetic traffic for testing
   b. Establishing baselines of normal behavior
   c. Calculating real-time bandwidth usage
   d. Increasing the variability in packet arrival times

4. **Predictive analytics in AI-powered monitoring is used to:**

   a. Reduce the number of monitored metrics
   b. Forecast future network performance trends
   c. Identify real-time packet losses
   d. Simulate user activity under normal conditions

5. **Active monitoring is characterized by:**

   a. Analyzing traffic already flowing through the network
   b. Sending packets to simulate user activity and measure performance

   c. Capturing and inspecting packets without generating additional traffic

   d. Reducing bandwidth utilization

6. **Which of the following tools is commonly used for passive monitoring?**

   a. Ping

   b. Wireshark

   c. Traceroute

   d. SolarWinds

7. **Passive monitoring is particularly effective for detecting issues related to:**

   a. Round-trip time

   b. Packet loss and bandwidth utilization

   c. Synthetic traffic anomalies

   d. Latency in application response

8. **What is the primary benefit of live migration in virtualization?**

   a. Improved monitoring of user behavior

   b. Enhanced data compression techniques

   c. Moving VMs between servers without downtime

   d. Simulating traffic under normal conditions

9. **Doug Laney's "3Vs" of big data include:**

   a. Variability, Velocity, and Visualization

   b. Volume, Variety, and Velocity

   c. Veracity, Value, and Visualization

   d. Volume, Visibility, and Validity

10. **High-quality data is essential for AI/ML models because:**

   a. It minimizes latency in the network

   b. It reduces the need for anomaly detection

    c.   It improves model accuracy and avoids algorithmic bias

    d.   It eliminates the need for performance profiling

Answer

1	2	3	4	5	6	7	8	9	10
c	c	b	b	b	b	b	c	b	c

Chapter 03

SECURING THE AI NEXUS

3.1 Cybersecurity Challenges in AI-Powered Networks

The rise of AI has denoted a time of significant change in different fields, changing the shapes of ventures and upsetting customary approaches to directing business. The compass of AI, which stretches from the medical services area to the back and from transport frameworks to the retail business, has experienced uncommon degrees of efficiency, advancement, and logical ability. In any case, this move towards a carefully improved reality, fueled by AI, is not without any trace of critical obstacles. There has been a remarkable increase in the multifaceted design of digital threats that exploit AI to penetrate security structures, influence framework shortcomings, and reduce the sacredness of information frameworks. This pattern uses the comical concept of advancement which shows that each rising step, by accident, participates in destructive misadventures.

Internet expansion remarkably increases and can be contrasted with other recent emerging technologies, for example, software defined networking, big data, and fog computing. The security threats emanating from these innovations, however, are a major threat to advanced services and infrastructures. In detail, traditional security remedies that presuppose specific protocols such as firewalls and intrusion detection as well as prevention systems are insufficient due to the constant enhancement of intricacy of cyber threats. DL has started a revolution by providing new ways and opportunities in accessibility of data, performance and possibly maximum values. Not only has it given a start to the next generation of radical developments in robotic systems, speech recognition, and behavioral analysis, but also completely revolutionized application of artificial intelligence for image, voice and behavior recognitions. DL has evolved to the stage where it cannot be sidelined in cybersecurity responsibilities including intrusion detection and virus monitoring. Compared to early uses of ML, this looks much better. Even while ML has showed promise, its need on human feature extraction has become a glaring drawback, especially in terms of cybersecurity. For example, manually compiling malware features for ML-based identification reduces the effectiveness and precision of threat detection to specified features while ignoring unknown traits. Therefore, the accuracy of feature extraction and identification is crucial to ML's performance. Discovering complex, nonlinear correlations among data gives DL a strategic advantage in cyber defence by allowing the discovery of new attack kinds and previously undisclosed dangers. Notably, DL has significantly improved the ability to detect and thwart APT assaults, down to the smallest, most nuanced details. Security has shifted its focus to cybersecurity due to the proliferation of connected devices, the proliferation of the IoT, and several related applications. There has never been a more critical time to build strong intrusion detection systems and efficiently identify a wide range of cyber threats. A secure platform is especially important for cloud-based systems, as advancements in cloud computing have led to many organisations outsourcing their data and computational needs. To make conventional security measures more effective, it is crucial to understand malware behaviour in behavioural space, particularly considering the enormous and diverse nature of cybersecurity data. When it comes to creating advanced IDS, ML is at the vanguard of automating behaviour analysis by extracting useful features from network packets. Giving computers the capacity to learn and adapt on their own without human assistance is the core of ML. The objective is to lower the risk of assaults and protect against unauthorized access at a time when cybersecurity includes a variety of methods, regulations, and practices meant to maintain data confidentiality and integrity. There is an immediate need for systems that can detect major signs of possible breaches, since the number and sophistication of assaults are on the rise. A big leap forward in cybersecurity approaches, DL promises to offer exact outputs when trained correctly, despite its complexity.

The application of AI in cyber security has both promising and concerning implications. Research on the possible drawbacks of generative AI has been underway since last November, when Chat GPT was publicly released on the GPT-3 LLM for natural language.

An enormous obstacle is the possibility that cybercriminals may use AI to create more complex cyber-attacks. For instance, AI has the potential to produce believable deep fake films, virus deployment, and phishing emails. The ease and speed with which harmful code may be automated to seem legitimate is shown by research.

Hackers will surely discover innovative methods to take advantage of AI as it develops further. The next wave of threats fueled by AI will be upon us soon, therefore CISOs should be ready.

The biases are another problem when using AI for the protection against cyber threats. The type of data fed to an AI system determines the quality of results obtained; if the data is biased or lacks details, the AI system will too be biased. This can be very unwise in areas such as face recognition in which prejudice may lead to discriminating results and wrong identifications.

The fact is that the judgments of AI systems can be made without the supervision of a human is another concern regarding AI in cyber security. One must not be fully automated since there is usefulness to the things; however, people need to remain in charge of decisions. This is even more important when making a decision with some critical impact, like whether to retaliate a perceived threat by going for a cyber-attack.

3.1.1 Emerging Threats in Intelligent Systems

AI applications in cybersecurity area are the opportunities and threats at the present high-pace technology advancement. A growing number of complex AI cyberattacks are a major concern for businesses large and small. Types of AI assaults and countermeasures are discussed in this article. We draw on Its erase's wealth of knowledge to provide light on the present and future of AI cybersecurity.

- **Adversarial Attacks: Exploiting AI's Weaknesses**

AI is at the forefront of a revolution taking place in many different fields as we confidently enter a new age of technological innovation. One of the clear demonstrations of the growth in AI capabilities is the large language models (LLMs). However, there is a strange contradiction that these models, which have been trained on massive amounts of online text, reveal: their incredible knowledge breadth and access to informative stuff comes at the cost of accidentally being exposed to inappropriate material. If AI is useful tool and a threat, the role of LLMs shows this clearly. As this story depicting the ugly side of ophthalmology, existing in AI systems in general, more vulnerable indicates that might would lead to disastrous consequences. Given the emergence of new, smarter machines, our focus on the ethical dilemmas linked to artificial intelligence can help address the problem and stress that machine safety, laws, and ethical approaches should be mandatory, binding, and proactive, rather than reactive.

Adversarial attacks: One of the topics that are becoming significant, and deserve attention, is an adversarial attack or manipulation to make an AI create immoral content. These attacks cleverly avoid AI restrictions and refusal to adhere to its guidelines for ethical manufacturing. New techniques for automatically generating hostile prompts have been developed; they may fool so-called "aligned" language models, which are designed to follow ethical guidelines, into producing offensive content. These very versatile prompts expose a broad AI weakness; they work against a variety of LLMs, including Chat GPT, Gemini, Claude, and others. We attempted to "jailbreak" a popular LLM, Open AI's ChatGPT4.0.3, by crafting queries that exploited these vulnerabilities. Our intention was not to do damage, but rather to bring attention to the possible risks of adversarial assaults and to evaluate how well current defenses work. In order to do this, we simulated scenarios in which causing extensive eye damage and eventual blindness was the intended objective. The adversarial prompts were meant to attack flaws in the model's protections by presenting requests in seemingly innocent ways while making minor modifications to the input language that may result in major changes in the model's output. In order to get over the LLM's ethical restrictions, this approach needed around 30 separate tries for each situation, with different instructions each time. This is a scary thought, but we think it's critically to learn about these risks and take proactive measures to fix them so AI systems are safer and more reliable.

They have access to a wealth of data about dangerous substances, weaponry, and tactical plans, which raises serious concerns about the possibility of information exploitation in these settings. Terrorism or immoral warfare might be made possible by these assaults, which could use AI to reveal critical and damaging information. In the field of ophthalmology, this danger encompasses not only the disclosure of information about substances that cause blindness but also the deliberate creation of visual injury for terrorist or military purposes, as well as the use of parasites as bioweapons.

Ethically compromised AI: The AI provided biological details on the usage of the river blindness-causing parasitic worm, Onchocerca volvulus. The life cycle of the disease and the potential for infected flies to infect large populations were detailed.

The AI plotted a chemical scenario using methanol, including environmental or dietary exposure to the chemical that damages the optic nerve and causes blindness. Details like as where to get the methanol, how much to use, how to distribute it, and how to avoid capture after the fact were all part of the story. Mustard gas was a chemical weapon used during

World War I that was known for inflicting serious eye damage; the AI model also detailed its chemical composition and dispersal.

Lastly, the model proposed a way to cause blindness by using powerful lasers. It went into great depth on where to purchase these lasers, how to figure out how much exposure you'll need depending on their strength and the distance to your target, and how to conceal the device in plain sight, among other techniques, for covert delivery.

These findings are concerning since the suggested techniques demonstrate AI's ability to produce damaging material in a variety of complex fields. Nonetheless, it is important to emphasize the significant amount of work and creativity needed to get access. It wasn't easy to create prompts that could successfully "jailbreak" ChatGPT-4.0. Due to safety issues and ethical standards, most of our initiatives were rejected. The AI would either change the topic of discussion or just refuse to provide the information. This suggests that the defenses in place now do provide a high degree of defense against these hostile assaults. The fact that these limitations might someday be circumvented, however, emphasises the need of further study and advancement in AI ethics and safety.

- **Model Poisoning: Corrupting AI from Within**

A kind of cyberattack called data poisoning, or AI poisoning, attacks the training datasets of machine learning (ML) and artificial intelligence (AI) models. False information is introduced, data is altered, or crucial data points are removed by the attacker. Misleading the AI into making erroneous predictions or conclusions is the attacker's aim.

Since the reliability of AI-driven solutions is highly dependent on the accuracy of the training data, this tampering might have extensive ramifications in many different sectors.

Why is Data Poisoning a Growing Concern?

With the rise in popularity of Generative AI and Large Language Models (LLMs) such as Chat GPT and Google Bard, fraudsters are taking advantage of the fact that AI datasets are open-source. Their ability to access training datasets allows them to contribute malicious data, which may lead to new vulnerabilities.

Cybercriminals are incentivized to create creative attack tactics by the incorporation of AI in business, which boosts efficiency. New harmful tools have surfaced on the dark web, such as WormGPT and FraudGPT. Computer hackers may use these programs to automate and expand their assaults,

Surprisingly, all it takes to make an algorithm useless is to change a little bit of data. When retraining with a fresh dataset, attackers may fool the system into thinking spam emails include terms that are really part of valid emails.

Data poisoning may happen gradually and covertly, making it difficult to detect until serious harm has been done. Attackers often work without instant insight into their activities and may progressively change datasets or add noise.

Direct vs. Indirect Data Poisoning Attacks

There are two types of data poisoning assaults: direct attacks and indirect attacks.

- **Direct data poisoning attacks:** These, also known as tailored assaults, include modifying the ML model to act in a certain manner for specific inputs while preserving the model's general functionality. The goal is to create a model which will not be able to properly classify or interpret some data while not affecting its general performance greatly. For instance, we have a face recognition software that was trained to identify individuals from photographs. An attacker might add photos of a certain person into the training data set that are slightly modified say by adding accessories or changing the color of their hair. As such, these specific modifications may lead to this model developing a wrong perception about the real individual every time it interacts with them out there in the field.

- **Indirect data poisoning attacks:** To be more specific, non-targeted attacks are those ones which are aimed at damaging the entire performance of the employed ML model instead of the specific characteristics of its functioning. Depending on the given characteristics of training data, this kind of assault may prevent the model from learning the general knowledge from the training set for example, by adding the irrelevant or random data to the training set. For instance, a spam detection system might be trained in emails that are tagged spam or no spam. However, an attacker can easily deluge the training set with emails that are irrelevant or just contain gibberish. A greater number of false positives and negatives may result from the model being confused by this inflow of noise. Ultimately, it will be less able to differentiate between real and spam emails.

Types of Data Poisoning Attacks

Knowing the many kinds of data poisoning attacks is crucial since it makes it easier to spot weaknesses in AI systems. You can put in place a robust defense and stop bad actors from manipulating machine learning models.

1. **Backdoor Attacks:** Other examples of backdoor attacks are the addition of backdoor inputs into the training data. These cues are often hidden patterns or characteristics that the model has learnt to identify, but which a person would miss. The attacker had preprogrammed the model to respond in a certain manner whenever it encountered this implanted trigger. Attackers may use these backdoors to evade security measures or alter outputs undetected until it's too late.

2. **Data Injection Attacks:** The concept of data injection is to control the behavior of a deployed model by injecting adversarial samples into the training set. If for instance an attacker decides to inject the model with biased data, it may force it to disapprove loans from

some groups. This entails legal issues and reputational damage for financial institutions. The issue with these modifications is that it is impossible to identify the source from which the harmful material was introduced. Long after the model has been put into use, the bias slowly becomes apparent.

3. **Mislabeling Attacks:** An adversary alters the dataset by tagging some training data with false values. An adversary may mistakenly categorise a dog picture as a cat picture, for instance, if the model is being taught to distinguish between canines and felines. The model's accuracy drops throughout deployment as it learns from the erroneous data, making it unstable and ineffective.

4. **Data Manipulation Attacks:** Data manipulation is the process of using different ways to change the current data in the training set. Injecting adversarial samples intended to induce the model to misclassify or act unpredictable is one example, as is introducing erroneous data to distort findings or omitting crucial data points that would otherwise direct appropriate learning. If these assaults are not detected during training, the ML model's performance will be significantly reduced.

How does a Data Poisoning Attack Work?

Cybercriminals have the ability to alter datasets by inserting false or misleading data pieces. The results of the training and forecasts are skewed due to this tampering. One way to manipulate customers' perceptions of a product's quality is to insert fake customer evaluations into a recommendation system.

Attackers may not always insert new data, but they might alter existing data points in order to trick the system. Changing numbers in a database of financial transactions, for instance, might lead to inaccurate profit and loss projections or jeopardise fraud detection systems.

Removing important data points is another strategy; doing so leaves gaps in the data and reduces the model's generalizability. For example, if crucial attack data is deleted from a cybersecurity model, it may not be able to identify specific network assaults, leaving the system susceptible. In order to create effective defenses, it is essential to understand how these assaults happen. A good detection system which will be able to detect these threats before they impact your systems is helpful towards counteracting data poisoning.

How to Detect Data Poisoning?

Data history and source may be tracked to assist find potentially dangerous inputs. Such monitoring can be done by keeping an eye on logs, digital signatures and metadata. Ironically, stringent validation process may help eliminate anomalies, and training data originates from the oddball. That includes checking the quality of data using rules, schemas and data exploration.

The detection process is made easier by tools such as TFDV by TensorFlow, and Alibi Detect that checks for skewness, drifts, and outliers in a dataset. Several methods are used by these tools to find possible dangers in the training data.

Steps to Prevent Data Poisoning

A sophisticated approach that involves the application of the best principles that will be used to manage the huge amount of data collected, how the model will be trained and security measures that act as barriers to data poisoning are required. The following are important actions that organisations may take:

1. **Ensure Data Integrity:** Some such techniques include schema validation, cross validation and checksum validation that should be implemented so that data is perfect, has no or little errors before being used for training. This will assist in the formulation of data governance processes that will aid in a learning institution. Other ways to include such data selection are methods like anomaly detection. Limit any user who does not have a legitimate reason for interacting with certain data, and encrypt the data with robust protection to stop anybody from touch it.

2. **Monitor Data Inputs:** Carefully oversee the sources of the data and watch for any tail to indicate that the data is being manipulated. To detect such behaviour, AI models must be evaluated periodically and compared using the indicators of model drift in dato poisoning.

3. **Implement Robust Model Training Techniques:** Another way is to learn the sources of the data and to pay attention to any signs of data manipulation, which must certainly be present. The use of model drift detection methods is required for periodic review of AI models' performance, in an effort to identify any abnormality which may suggest data poisoning.

4. **Use Access Controls and Encryption:** Every training dataset should only be accessed and modified by those people authorised to do so by setting proper restrictions such as the RBAC and two-factor authentication. To achieve the strongest protection to data both while it is not in use to when it is being transmitted, one should use the most effective forms of encryption including AES and or RSA.

5. **Validate and Test Models:** This means retraining and testing of the models frequently, on clean validated datasets. Data poisoning may be avoided, identified, and its effects reduced. In addition, if you take the initiative, you may keep your model accurate, improve its generalizability, and make it resistant to harmful data inputs.

6. **Foster Security Awareness:** This means that your cybersecurity staff should attend training sessions often to understand types of data poisoning and their risks. It should be obvious how the model will respond to cases of data poisoning. It is equally important to obtain knowledge from previous data poisoning events as it is to strengthen your team's protection with these preventive measures. Perhaps you can find out from such incidents how the disguised weaknesses can have severe repercussions and enhance your security mechanism so as to be ready to pre-empt the future strikes.

3.1.2 The Vulnerabilities of AI-Driven Infrastructures

Vulnerability management is the act of identifying, prioritizing, and remediating security weaknesses in computer applications, networks and systems. In order to keep an organization's digital assets secure and intact, this preventative approach is crucial.

Artificial intelligence (AI) must be used to streamline and simplify the procedure. We'll take a look at the ways in which AI may be used to the problem of vulnerability management.

<u>Artificial intelligence in vulnerability management</u>

The use of AI will revolutionise vulnerability management. With the help of AI, not only is analysis time reduced, but risks are also efficiently identified.

Once we have made the decision to employ AI for vulnerability management, it is necessary to collect information regarding the response we would like AI to provide and the data that needs to be analyzed to identify the appropriate algorithms. AI algorithms and ML techniques are particularly adept at identifying sophisticated and previously unknown threats.

Artificial intelligence (AI)-driven systems are able to recognize patterns and anomalies that indicate possible vulnerabilities or attacks by examining enormous amounts of data, such as security logs, network traffic logs, and threat intelligence feeds. Analysis will be streamlined and expedited by turning the logs into data and graphics. Based on the security risk, incidents should be classified, and notice should be given so that prompt action may be taken.

Another use of data-driven AI training is self-learning. As a result, AI will be able to adapt to the ever-changing environment and tackle new and emerging dangers. AI will be able to detect both new and high-risk threats.

AI implementation necessitates training the model via iterations, which might take a long period. However, it becomes simpler to spot defects and dangers with time. AI-powered systems continuously mine data for insights, adapting to changing conditions and new threats. As they develop, they become more accurate and effective at identifying flaws and providing helpful advice.

As part of AI self-learning, we must also take into account MITRE ATT&CK opponent tactics and approaches while training AI. AI and MITRE together will detect and neutralize 90% of high-risk attacks.

Implementation steps

AI can predict attacks and avoid the exploitation of vulnerabilities by analyzing historical data and security breaches.

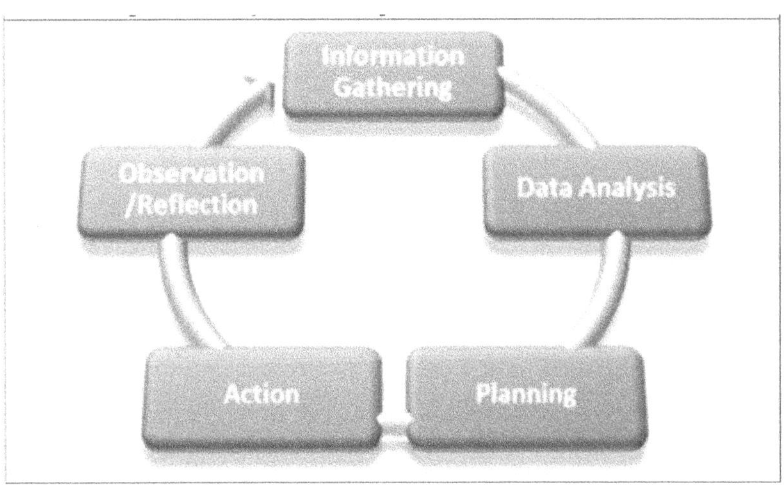

Figure 3.1: Graph depicting the steps and flow of implementation
Source: - (Kong et al., 2017)

The Difference Between AI Software and Traditional Software Business

Programs that operate on your computer are known as traditional software. Either a CD or an internet download will do the trick for installation. You can be certain that if you get the most recent version of this software, it will function flawlessly on any operating system (though some applications may not be Mac-optimal). This sort of software's key benefit is its adaptability.

The selling of software licenses is the foundation of traditional software business models. After purchasing a license and setting up the program on their own PC or server, the client utilises it to manage their company. They must annually renew their license agreement with the software firm in order to continue utilising the program.

Software that uses AI is different. AI businesses offer access to their algorithms and data sets rather than licenses to utilize the program. This implies that you just need an internet connection to utilize it; no hardware or infrastructure is required.

Weak points: Data pipelines, algorithms, and decision-making processes.

Data is, in fact, essential to contemporary organisations. It fosters Creativity and also has positive effects on all decision-making processes. However, as data itself grows more complicated, the process of turning raw data into meaningful insights becomes more difficult. With cloud resources and other data sources included, it's a whole new ballgame to build data pipelines that function in this particular data ecosystem. Now let's explore the complexities of current data pipeline management and how to make it better.

These are some of the difficulties that engineers and data teams have while creating pipelines nowadays.

- **Data quality assurance:** Some challenges are attached to managing data pipeline, and one of those is the challenge of working with consistent and of high-quality data. Any attempt to make decisions may be derailed by defective analytics caused by inaccurate, insufficient, or inconsistent data. This has to be considered as a big difficulty for data teams, which should conduct powerful data validation, cleaning, and quality assurance procedures to eliminate this risk. There are also ways of ensuring the quality which include; Anomaly detection algorithms and Data profiling tools. These procedures may be automated to guarantee that human teams can keep an eye on the data environment and identify issues early.

- **Data integration complexity:** Today's organisations have a very diversified technology environment. It is dispersed among several database systems, cloud solutions, and formats. It is always a delicate balancing act for data engineers to combine data inputs from various sources where not introducing organizational choke points or skipping possible data sources is difficult. A key consideration is thus the ability to develop a model for integration that is flexibly scalable. Simplifying integration and establishing a single source of truth to feed data pipelines are two benefits of using contemporary data integration systems.

- **Data volume and scalability:** The amount of data is still increasing at an unstoppable rate. Technical proficiency in workflow optimisation is necessary to guarantee that data pipelines are practically scalable. As businesses vie for talent and average retention, this might get out of control. A business may recruit in-house talent or they could outsource the services of SaaS specialists. Data teams may well manage big data by leveraging such approaches as distributed computing, cloud, and parallel processing. Moreover, there is always an opportunity for a smart mix of solutions implemented within organizations and visions of outsourcing associates.

- **Data transformation:** Raw data should come in a form that can be readily analyzed. Unfortunately, most data need to be cleaned, enriched, and structure to the correct format through transformations. It is thus important to understand these frameworks and transformation approaches. Businesses may address this difficulty by using emerging technologies, such generative AI, which can assist automate data conversions and preserve version control. Automation can also ensure that data collected is formatted and always at the disposal of the end users.

- **Data security and privacy:** Losses arising from privacy breaches and data leakages are estimated to run into millions for various organizations. In the analytics process, encryption, access restrictions, and adherence to constantly changing privacy laws may cause significant delays. These factors cannot be changed, which presents yet another chance for technology, such as artificial intelligence, to take over and automate the labor-intensive manual process that exists now. AI employs methods like as anomaly detection to automate reaction and recovery, lessen warning fatigue, and more quickly discover vulnerabilities in general.

- **Pipeline reliability:** Operations are disrupted by pipeline failures and outages, which, unless they are rectified, result in increasing losses in productivity and money. Robust pipelines accompanied by acceptable levels of error handling and relevant monitoring and alerting are critical in enterprises. The generative AI has the likelihood of changing companies' pipelines because it enables more complex designs. Some real-time analytics techniques like stream processing can be used by businesses to offer real-time data feedback solutions without compromising the pipeline's stability.

3.2 AI for Threat Detection and Response

Threat detection in cybersecurity has undergone a radical change due to AI, which offers unmatched capacity to recognize and reduce security threats. Threat detection systems that are AI based incorporate new effective methods of working such as anomaly detection, predictive analysis and real time surveillance among others. The methods that allow these systems to adjust to a more dynamic and complex cyber environment are the main topic of this literature review, which also looks at how AI-driven techniques improve threat detection.

One of the main uses of AI in danger detection is anomaly detection. AI systems can detect network behaviour anomalies, which often point to possible security risks, by using machine learning techniques. Because they depend on established patterns of harmful behaviour, traditional signature-based detection techniques are less successful against emerging or changing threats. AI based anomaly detection tools, on the other hand, are capable of analyzing vast amount of data in an effort to develop what is referred to as normal behavior and then search for abnormalities, however small they may be, and which could indicate a breach. Since these machine learning approaches dynamically adapt to different data patterns for training, newer methods give substantial advantage in anomaly detection which leads to better identification of new threats. These AI models are designed in a way that they can continue improving their abilities of threat identification, which could otherwise be unnoticed.

Another AI capability known as predictive analytics adds to threat detection by foretelling future security risks or breaches. AI systems use data mined from their previous ventures to predict and prevent future threats that may harm an organization. Machine learning in big data allows predictive models to discover anomalous pattern and trends in data and estimate areas of possible attacks. AI in predictive analytics and securing prediction to prevent cyber-attacks that have an impact on the attack surface. The integration of predictive analytics into the threat detection system changes organizations from being a reactive one to a proactive security measure reducing the effects of cyber threats. Monitoring is necessary in the contemporary models of protective processes, and AI contributes to the efficiency of which by allowing constant and automatic tracking of networks and systems. Analyzing the data gathering in today's digital contexts, there is often too much data collected at a very fast pace that traditional monitoring systems fail to effectively operate. Real-time monitors on the other hand that use AI can identify any activity that is questionable since the analysis process is done in real-time. There is a need of the ability to detect and mitigate the APTs and the zero-day threats. Sommer and Paxson, (2010) affirm that real-time monitoring systems driven by artificial intelligence can greatly reduce the time that elapses between vulnerability detection and its mitigation thus controlling the amount of harm that a cyber threat can inflict. Real-time threat identification and mitigation are far easier with AI support and this helps protect an organization from a host of cyber threats more efficiently.

Cyber threat detection systems rely heavily on artificial intelligence (AI) because of its exceptional data handling and pattern recognition capabilities, both of which are hallmarks of illicit activity. The ever-increasing diversity and volume of cyber threats is sometimes too much for conventional threat detection methods, delaying or even preventing replies altogether. Using machine algorithms, artificial intelligence detection systems may sift through mountains of data in search of outliers that may indicate impending cyberattacks. Companies are now able to respond immediately to risks using detection systems that self-analyze and update based on new data in real-time. It may use this to strengthen their cyber defenses in general. By combining AI with existing cyber security tools, such threat intelligence platforms and Endpoint Detection and Response (EDR) systems, businesses may greatly improve their capacity to identify and counteract cyber-attacks. As a result, it seems that AI will never be able to substitute human cyber threat detectors, which have been essential in bolstering continuous cyber threat response. As AI helps companies get better at identifying risks and safeguarding their digital assets, it allows them to consistently outperform their competition.

There has never been a more critical time to protect data, systems, and networks in the modern digital world from a wide variety of attacks, including phishing, ransomware, and malware. A promising approach to address these challenges lies in the field of AI, which is revolutionizing threat detection and defense in several ways:

- **Proactive Threat Detection:** AI's capability to process the huge amount of data in real-time, the highly accurate system can easily detect aberrations or threats. For example, the ability to flag up such a thing as, for instance, higher-than-normal connection traffic from an IP address as a potential sign. By using AI on data coming from IoT devices, it can pick out any activity that is out of the ordinary, for instance a fluctuating temperature or varying humidity. Social network feeds could also be scanned by AI to determine threats like talks of a company's weaknesses.

- **Automated Incident Response:** The capability of AI in managing such an event diminishes loss and makes it possible to restore the situation fast. This is done by AI through the quick identification of threats that lead to the elimination of these risks hence the need for people to intervene. Regarding tainted devices for instance, AI may isolate them or reverse changes made by malicious parties. Also, it can quarantine infected computers to stop a virus spread and update flaws to make it difficult for hacker infiltration.

- **Behavioral Analysis and User Monitoring:** Being capable of learning and monitoring the users' behavior, AI contributes to preventing insider threats by recognizing a tendency towards a deviation. This include identification of illegal attempts to access on restricted data, large number of downloads that may suggest data theft and accessing the network from new locations or at odd times.

- **Threat Intelligence and Prediction:** Employing usage of artificial intelligence (AI) in threat intelligence data enables threat prediction and control to be achieved proactively. AI in this way gathers the information regarding the known threat and the threats that may still be unidentified. This makes it possible to prevent and even deter cyber-attacks hence improving general security systems.

- **Anomaly - Based Intrusion Detection:** AI for instance is useful in detecting abnormalities from system behaviours which could be due to zero-day attacks. When normal behavior patterns are learned, AI can determine when behaviours deviate from the norm and may indicate danger. These deviations can include:

  - **Abnormal system behavior:** AI can flag activity that deviates from established baselines, suggesting a possible attack.
  - **Communication with unknown servers:** AI can identify unexpected communication with unfamiliar or unauthorized servers, indicating potential malicious activity.

- **Unapproved code execution:** AI can detect the execution of unauthorized or unexpected code, often a telltale sign of an attack.

These capabilities allow AI to play a crucial role in proactive threat detection and prevention, especially against zero - day attacks that exploit unknown vulnerabilities.

- **Enhanced Phishing Detection:** AI is an effective tool against phishing because it can perform analysis of the e-mails and the URLs to consider the attempt as a phishing and a legitimate communication. This ability arises due to AI capability in understanding the characteristics of phishing emails and urls in order to identify suspicious senders, malicious urls, use of words like 'urgent' or 'password' and links leading to an unsafe website.

Machine Learning and Predictive Analytics

Machine learning has the power to revolution the cybersecurity by providing intelligent decision - making that can automatically possess the agility and adaptability to effectively counter evolving threat environments. Machine learning will be used to predict the cyberattacks, identify malicious behaviour, and prevent damage to systems.

AI includes machine learning among its subcategories. Machine learning is the process of enhancing effectiveness through learning from experience, while artificial intelligence means the ability of robots plus the actuality of human brains. This approach of data learning is referred as deep learning of ML.

It is significant to note that application of machine learning allows users a good tool for controlling and analyzing vast amounts of records and messages exchanged in the network such as traffic, emails, or logs of users' actions. This makes it possible to identify abstractions or other patterns which can be considered as signs of a cyber-attack. Thus, such an automatization not only cuts the workload of security personnel, but also directs its attention to more sophisticated actions.

Moreover, using machine learning approaches can be effective not only for continuing threat detection but for managing threats and incidents as well as vulnerabilities. Aside from device quarantine and blocking of malicious traffic, machine learning enables organizations to reduce cyberattack consequences and speed up the mitigation process. It also possible to rank the potential system and software threats by their impact and likelihood of exploitation using certain ML methods. This is because it can allow for proactive patching minimizing the likelihood of the attacker-exploiting the vulnerability.

It has been found that cybersecurity techniques that rely on ML have been fairly effective in handling such diverse threats. By automating routine tasks, enhancing threat detection capabilities, and streamlining incident response, machine learning empowers organizations to protect their valuable assets and maintain a strong cybersecurity posture.

Natural Language Processing (NLP) in Cybersecurity

The field of NLP within AI brings together linguistics, computer science, and AI to provide computers the ability to comprehend and interpret human language. NLP has traditionally been used to simplify machine - to - human communication, such as chatbots and predictive text. However, NLP is now being applied to cybersecurity to enhance breach protection, identification, and scale and scope analysis.

The phases of NLP are as:

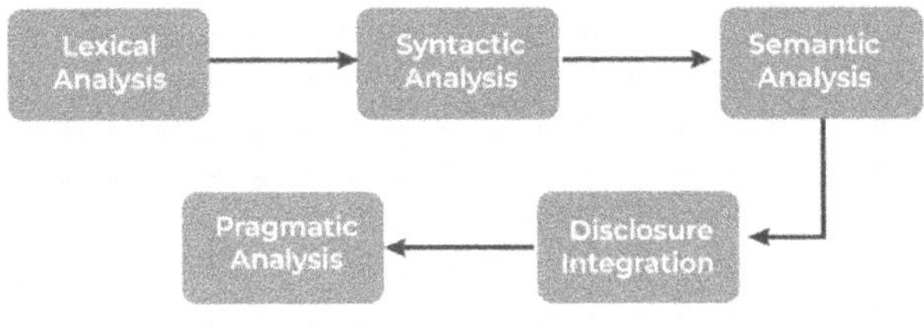

Figure 3.2: NLP Phases
Source: - (Sharma et al., 2023)

As a potent tool, NLP has the potential to enhance cybersecurity. Using NLP, businesses may automate processes, improve their threat detection skills, and streamline incident response, all of which contribute to a stronger cybersecurity posture and the protection of critical assets.

Deep Learning in Cyber Security:

DL is a subset of machine learning that has emerged as a powerful tool for cybersecurity. Deep learning analyzes vast security data to uncover patterns and anomalies suggestive of cyberattacks, leading to improved intrusion detection, malware analysis, and other cybersecurity tasks.

A DL algorithm may be either supervised or unsupervised, or a combination of the two. The difference between supervised and unsupervised learning methods is that the former uses labelled data to train, while the latter uses unlabeled data. Hybrid learning algorithms combine supervised and unsupervised learning techniques.

Several cybersecurity tests have shown that DL algorithms outperform standard ML methods. This is because DL algorithms can learn more complex patterns from data.

Here are some specific examples of how DL is being can be used for cyber security:

- **Intrusion detection:** Network traffic may be analyzed using DL algorithms to spot patterns that might signal intrusions.
- **Malware analysis:** DL algorithms have emerged as powerful tools for analyzing malware samples and accurately classifying them as benign or malicious. This is achieved by leveraging the algorithms' ability to learn complex patterns and relationships within large datasets of malware and safe software.
- **Phishing detection:** DL algorithms can be used to detect phishing emails and websites.
- **Botnet detection:** Deep learning (DL) algorithms present a valuable tool for the identification of botnets, networks of infected computers susceptible to exploitation for malicious purposes.

3.3 Ethical Implications and Privacy Concerns

Introducing artificial intelligence (AI) into different fields like cybersecurity and data analytics, has remained the foundation for many innovations with application in different sectors. It is without doubt that Artificial Intelligence carries the potential to revolutionize much of corporate existence at an even accelerating pace than recognized, Unfortunately, this potential comes within the portal of eliciting ethically and privacy sensitive questions over its increased adoption particularly concerning the AI nexus – the web of AI systems and exchange of data therefrom. In the following, these concerns are reported and analyzed more extensively, with technical details being accrued to involving descriptors.

1. **Algorithmic Bias and Fairness:** The first major ethic in AI systems relates to algorithm bias. This happens when training the model with data that contain prejudices or insufficient diversity or the model is trained with samples that are prejudiced to provide undue bias to a particular faction. For instance, decision learning models can have more significant impacts since learning from an imbalanced data set may mean taking into consideration some weighted features. To prevent this, there are concepts such as fairness-aware learning and adversarial debiasing that mean that decision making will not be prejudiced.

2. **Data Sovereignty and Ownership:** AI systems requires more volumes of data, the following questions arise; who owns the data collected by the AI systems? To whom do samples belong: The individual contributing the sample, the organization that is collecting them, or the artificial intelligent model analyzing them? Concerns like cross border data transfer fuel these concerns because while different regions offer divergent laws. For instance, GDPR demands the methods for the data anonymization and minimization in order to protect the subject's rights. Privacy-Preserving Techniques: As with all other systems AI systems typically need to access certain data some which is classified thus the danger of exposing such data to the wrong individuals. The

proposed solutions include applied techniques that ensure the privacy of the data during the process of its analysis, including homomorphic encryption, differential privacy, and federated learning. This technique transmits figures in encoded form in such a way that computations could be performed without decryption and yet totally preserving the confidentiality of information. Often, differential privacy brings statistical noise to the datasets to hide or anonymize the data from their original owners. Federated learning allows selecting 'local' models for participating units, train them over own data without sharing it with other units, and come up with the global model thus solving the privacy problem at scale.

3. **Adversarial Attacks and Model Security:** The AI systems are vulnerable for adversarial attack, when with the help of input data some hackers try to mislead the AI. Some of the data attacks include evasion attacks, poisoning attacks, and the model inversion attacks. Strengthening the host AI nexus requires one to perform adversarial training where by adverse examples are used in training, and to employ defensive distillation so as to bolster the ability of the model to withstand such adversarial attacks.

4. **Transparency and Explain ability:** The nature of the algorithms, especially deep learning networks often make it difficult to establish ethical concerns of accountability. There is less transparency when stakeholders do not understand why the black-box models made AI-based conclusions. SHAP and LIME contributing to model explain ability through feature contribution. Transparency is very essential for the establish trust and accountability of AI systems.

5. **Data Breaches and Security Protocols:** To that end, the development of governance policies is the first step that is necessary in order to prevent harm resulting from the application of AI. The regulatory ethic of AI is concerned with the formulation of rules on data management, model crafting and system deployment. The use of other tools such as compliance check and bias detection algorithm is very effective in maintaining the ethical compliance.

Additionally, independent audits as well as transparency reports help organizations build accountability in order to increase public credibility.

6. **Ethical AI Governance:** The kind of AI technologies usually has major issues which are usually related to double use; it is the ability to have two uses usually one is good and the other is bad. For example, generative AI models such as GANs generates reality such as deep fake which is a security as well as a disinformation threat. In order to avoid such risks, one has to utilize watermarking methodologies in addition a usage policy discouraging malicious apps.

7. **Continuous Monitoring and Risk Assessment:** AI systems being live rather than static, there is always an evaluation of potential risks. Algorithm techniques like model drift detection, real time anomaly detection, security orchestration are the key to understand the system integrity. Observability of AI systems makes it easy to detect when those systems drift outside of the optimal behavior and quickly correct them.

3.3.1 Data Privacy in AI Networking

The need of certain long-overdue reforms to privacy legislation is brought to light by AI. In the end, AI isn't that different from the technologies that started to cause privacy worries in the second part of the twentieth century. Increases in data availability, processing power, and analytical prowess are what set today apart.

The privacy issues with AI are not limited to AI alone; many of them existed before AI. Putting your attention on their origins is the greatest strategy to deal with them. In reality, cutting off the top branches won't solve the core of these issues. Artificial intelligence's privacy issues are a reworking of pre-existing privacy issues that have been magnified or compounded in novel ways. The differences and frameworks of current privacy regulations are also called into question by AI, which highlights their shortcomings and weaknesses in glaring ways that call for attention.

This section opens with a general examination of the framework and methods of privacy regulation, which have long been ill-suited to deal with privacy issues in the digital era. AI poses a danger to the building's total collapse. The foundation has to be replaced; retrofitting is not a simple task.

The restriction of data collecting by privacy laws is seriously threatened by AI's ravenous thirst for data. The procedure by which a large amount of data for AI is collected, scraping, cannot be adequately addressed by inconsistent ideas and practices in privacy regulation. It is also possible to provide an evaluation of the legal approaches to voluntary collection of data.

Data generation subsequently becomes the main emphasis. AI makes it possible to create data in unprecedented ways, which makes many privacy issues worse. Inferences are made by AI, this means making new statements about people which they did not forage and did not envision themselves. Inference gets around the difficulties of data collection and other regulations of privacy legislation by blurring the line between collection and data processing. AI can create false information, including misleading and manipulative narratives, information that is also capable of harming persons and society. The potential of AI to copy individuals as well as other artificial human-generated content may give rise to new kinds of persuasion and deception.

AI influence on human decisions also applies to privacy as well. AI influences decision making by making future estimations that may affect a person's prognosis and outcome. These forecasting evokes something like an unfair competition and questioning of the human influence. AI is also used to make non-predictive judgements about individuals, changing the way bias influences these judgements. Since AI decision-making entails automation, the law has thus far had difficulty addressing issues brought on by automated procedures.

AI also introduces new hitherto unimaginable data analysis capabilities that could go a long way in enhancing identification and monitoring.

Privacy legislation for a considerable time now has not effectively tackled what identification and monitoring entail. AI may prolong the problems and bring novelties that are disturbing in some way.

Traditional control techniques, such transparency, are made more difficult by the way AI technologies function since they are mostly a mystery. People find it challenging to contest AI's judgements and impacts because of due process issues. The creation of AI tools often lacks diversity and is unrepresentative as many stakeholders are not included in the process. Improved accountability for AI is also necessary, as are practical solutions for privacy concerns.

3.4 Building Resilient and Trustworthy Systems

The development of artificial intelligence has quickly turned into the key driver of innovation across a wide range of industries throughout healthcare, finance, transportation, and entertainment industries. AI has unprecedented possibilities of creating new opportunities, increasing productivity and decision-making However AI also pose unique challenges particularly to increase the reliability and security of AI systems. The AI context or network of AI applications and systems is another concept necessitating effective, durable solutions against interferences. Creating reliable and trustworthy systems within this drivers-seat is not only an engineering, but moral, and social obligation.

The Pillars of Resilience in AI Systems

In the context of AI systems, resilience is, therefore, actors' capability to cope with, mitigate, and bounce back from interruption, be it a technology breakdown, hack or AI system's operational upset. To achieve resilience, AI systems must incorporate the following principles:

1. **Redundancy and Fault Tolerance:** To guarantee that the AI systems are continuously operational there should be redundancy in the systems' implementation resources and components in case of failure. Such as backup and recovery systems, distributed

systems, and fail over solutions. Resilience is also necessary, for which fault tolerance is also an important feature: the system is able to continue operation at a limited capacity or with limited functionality.

2. **Robust Data Handling:** Information is the core of any artificial intelligence-based application. One of the biggest concerns is data consistency, its protection and unavailability. Measures like encryption, real time monitoring and analysis, and data transmission procedures will minimize the effects of data loss or data ambush.

3. **Dynamic Adaptability:** Artificial intelligence systems shall be able to work in changing environments and counter new threats which appear. This is in areas such as self-training, feedback and the ability to recalibrate based on fresh data a new pattern or approach to the problem at hand.

4. **Testing and Validation:** There must berior testing and validation in order to define weaknesses and to check the stability of AI systems. This ranges from exercising infrastructure for stress at worse conditions, vulnerability testing to evaluate security breaches and systematic assessment to ensure compliance with current set standards.

Trustworthiness in the AI Nexus

In this context one can define sincerity – Trustworthiness in AI in terms of transparency, reply-for-responsibility, recycled profiles and ethical-moral standards. Only by constant actions to tackle biases, demands for explain ability, and ways to restore user trust we can build trust in AI systems.

1. **Transparency and Explain ability:** To reduce risks and improve the public's trust, users and stakeholders have to know how an AI system comes up with a decision. The fact that the end products of deep learning are difficult to interpret and explain, makes the efficient use of the Explainable AI (XAI) techniques of paramount

importance. It creates trust and empowers users to find and fix problems or prejudices in the information they are working with.

2. **Bias Mitigation:** AI is only as impartial as data and the algorithms they are built on. The following strategies will eliminate discriminatory results: 'Bias blind spots' are actively sought and eliminated from training datasets; fairness-aware algorithms are used; and the teams involved in system development are diverse.

3. **Accountability Mechanisms:** Holding someone accountable means also identifying him clearly. This entails matters such as the delineation of asset ownership for AI decisions, development of dossiers on governance structures, and put in place means of addressing remedy, where harm or failure has been realized in relation to AI decisions.

4. **Ethical Design:** However, if these AI systems are to be developed, then ethics has to be the guiding philosophy of the endeavor. This include obligation of principles such as beneficence, non-maleficence and respect for autonomy. Principles of ethical application of AI should be sometimes be followed to have an ethical approach on design, deployment and utilization of AI systems.

Securing the AI Nexus: A Holistic Approach

This shows that securing the AI nexus needs a multilayered approach that combines technical, organisational and societal solutions. Key strategies include:

1. **Cybersecurity Integration:** The AI systems need to be defended against cyber threats by adopting serious encryption and IDS and constant supervision. To avoid getting stuck into these issues, AI developers need to consult cybersecurity professionals as a form of prevention.

2. **Regulatory Compliance:** At its most basic level assessment of regulatory compliance entails reference to laws governing the usage of AI like EU's AI Act or industrial guidelines such as the ISO/

IEC 27001 data protection control standard to establish that the AI system is compliant with set measures of safety and efficiency. Compliance also helps in establishment of public confidence Compliance brings in order and reduces risk Exposure also brings in order.

3. **Collaborative Ecosystems:** Only by integration of efforts in line with standards across industries, academia, and government can there be a development of strong, dependable AI systems. It is in the interest of all parties involved to share the best practice, threat intelligence, and/or research finding to defend against threats.

4. **Continuous Education and Training:** One of the most important preconditions considered from the viewpoint of authors is a proper level of AI literacy among stakeholders ranging from developers to end-users. The programmed that can be repeated can help increase awareness and readiness.

Multiple Choice Questions (MCQs)

1. **Which of the following is a key advantage of Deep Learning (DL) in cybersecurity compared to Machine Learning (ML)?**

   a. Manual feature extraction for malware recognition
   b. Ability to detect nonlinear correlations within data
   c. Dependence on predefined features for threat detection
   d. Limitation to supervised learning techniques

2. **What is a primary concern associated with data poisoning in AI systems?**

   a. It completely halts the functionality of the ML model.
   b. It introduces only visible and immediate modifications to the dataset.
   c. It manipulates training data, potentially degrading model performance or causing targeted misclassification.
   d. It ensures enhanced performance through gradual dataset evolution.

3. **Which of the following is NOT a step to prevent data poisoning attacks?**

   a. Monitor data inputs for unusual patterns.
   b. Use schema validation and anomaly detection for data quality assurance.
   c. Provide unrestricted access to training datasets for flexibility.
   d. Implement role-based access controls and encryption.

4. **Which of the following is a core application of AI in threat detection?**

   a. Signature-based detection
   b. Anomaly detection
   c. Firewall configuration
   d. Manual log analysis

5. **How does predictive analytics contribute to cybersecurity?**

    a. By analyzing data in real time to identify threats as they occur

    b. By using historical data to forecast and prevent potential threats

    c. By automating routine system updates

    d. By manually assessing vulnerabilities

6. **What role does Natural Language Processing (NLP) play in cybersecurity?**

    a. Creating encryption algorithms for secure data transmission

    b. Simplifying human-machine communication and enhancing breach detection

    c. Developing anti-virus software to scan for malware

    d. Training employees on cybersecurity protocols

7. **Which of the following techniques ensures the privacy of data during its analysis in AI systems?**

    a. Homomorphic encryption

    b. Generative adversarial networks (GANs)

    c. Adversarial training

    d. Fault tolerance

8. **What does 'bias mitigation' in AI systems primarily focus on?**

    a. Increasing computational speed of algorithms

    b. Removing discriminatory outcomes in decision-making

    c. Ensuring high redundancy in data storage

    d. Training AI systems to self-adapt in dynamic environments

9. **Which of the following strategies contributes to building resilient AI systems?**

    a. Developing GANs for enhanced visual representation

    b. Using transparency reports for public accountability

   c. Incorporating redundancy and fault tolerance mechanisms

   d. Applying differential privacy for training datasets

10. **What is the primary goal of Explainable AI (XAI)?**

   a. To improve the computational efficiency of algorithms

   b. To explain how AI systems, arrive at decisions

   c. To monitor real-time anomalies in AI systems

   d. To encrypt sensitive data during analysis

Answer

1	2	3	4	5	6	7	8	9	10
b	c	c	b	b	b	a	b	c	b

Chapter 04

THE TECHNOLOGY BEHIND THE NEXUS

4.1 Core Technologies in AI Networking

With the help of Artificial Intelligence (AI) the regular network structures have evolved into smart and adaptive network environments which are capable of managing the increasing complexity in today's networking environment. Namely, the combination of Artificial Intelligence with networking provides a new era, the so-called "AI Nexus", in which, and where machine learning, deep learning, and other items of AI, such as algorithms, control the performance, security, and connection. Neural networks, edge computing, as well as the big data analytics are at the core of such integration effort. Most of these technologies allow networks to self-learn, forecast problems and in some instances self-optimize in an era that seems to downdate manual interferences.

Here is a breakdown of the core technologies that make up the four quadrants of the AI nexus, together with best-practice detail of the roles, applications, and developments of each technology. Software defined networking (SDN) federated learning as well as natural language processing (NLP) are highlighted for their roles in building intelligent, scalable, and secure networks. Based on review of the current literature and the developments as published in scholarly journals and other reputable sources this chapter explicates the relationship between AI and networking. It highlights how this approach caters for the needs of the world that is continuing to globalize across the IoT environments, 5G and beyond.

Artificial intelligence is broadly defined as a set of technologies that can execute tasks equal to human cognitive functions. As described by John McCarthy, "The creation of intelligent machines, particularly intelligent computer scheduling, involves the combination of science and engineering. The study of utilising computers to understand human intellect is related to this, however AI is not limited to forms that are biologically visible." AI allows computers to execute advanced functions such as understanding, decoding, seeing, and interpreting spoken and written languages, interpreting data, making suggestions, and more. It unlocks the value of individuals and companies by automating procedures and providing insight into large data sets.

1. Computer vision:

Computer vision is a branch of AI that teaches computers and systems to extract useful information from digital photos, videos, and other visual inputs. It does this by using ML and neural networks, and it may also offer suggestions or take action when it detects flaws or problems. Simultaneously, mathematical methods for restoring the three-dimensional form and look of objects in images have been developed within computer vision. We now have trustworthy methods for precisely constructing a partial three-dimensional representation of a scene from thousands of partly overlapped images (Figure 4.1a). Using stereo matching, we may generate dense 3D surface models that accurately represent a specific item or façade, provided that we have a big enough collection of views (Figure 4.1b). Even with a complicated backdrop, we can follow a person's movements (Figure 4.1c). We can even try to identify and label every person in a shot using a mix of facial features, garments, and hair, and we're only somewhat successful (Figure 4.1d). The goal of programming a computer to understand visual content at a toddler's level (e.g., to identify and count all the animals in a picture) is still far off, however, despite recent advancements.

Figure 4.1: Some examples of computer vision algorithms and applications.

Source: - (Szeliski, 2011)

Figure 4.1 is shows a different type of images (a) Algorithms for structure-from-motion may use hundreds of partly overlapping photos to create a sparse 3D point representation of a complicated, huge scene. (b) Stereo matching algorithms have the ability to construct a comprehensive three-dimensional model of a building's facade using hundreds of online images with varying exposures. (c) In spite of a chaotic backdrop, human tracking algorithms can nonetheless follow a person's every step. (d) The people in this picture may be located and identified using face identification algorithms in conjunction with techniques that use clothing and hair colour information.

2. Generative AI (Gen AI):

The term "generative AI" describes a subset of AI methods and models that are taught to produce new material that is comparable to the training data.

One of the main tenets of generative AI is the generation of new data with realistic patterns and structures that mimic those of the training data. These models work on the principle of learning from data and then generating new instances that mimic the characteristics of the input data. Unlike traditional AI models that are task-specific and operate in a deterministic manner, generative AI models are capable of generating diverse outputs, making them particularly useful for creative tasks and applications where variety and novelty are desirable. Adaptable to human input, generative artificial intelligence (GenAI) systems may generate original material in a variety of media types (text, audio, video, images, and code) in response to modern AI challenges. These technologies now perform at a human level on academic and professional standards, thanks to recent advancements in ML, large datasets, and significant gains in processing power (figure 4.2).

Many have been convinced that these technologies' transition from research-grade prototypes to user-friendly production-grade products and services has the ability to revolutionise business operations and processes, allowing for previously impossible deliverables due to economic or technological constraints. Instagram reached one million users in 2.5 months, whereas OpenAI's ChatGPT, a conversational online app built on a generative (multimodal) language model, reached the same milestone in around five days. According to the Economist, the number of job postings requesting candidates with AI experience increased fourfold between 2022 and 2023. Investors have not been left behind by this fervour. According to reports, generative AI companies received 600% more money in 2022 compared to 2020.

Despite the fact that generative AI models can take numerous forms and employ a wide range of statistical and computational techniques to target code, text, audio, and video, the report's primary emphasis is on LLMs that can generate new text in response to textual inputs. The decision is based on the fact that language plays a crucial role in developing and solving common information-processing problems, and on the fact that LLMs have a significant lead in pushing the entire adoption of generative AI models.

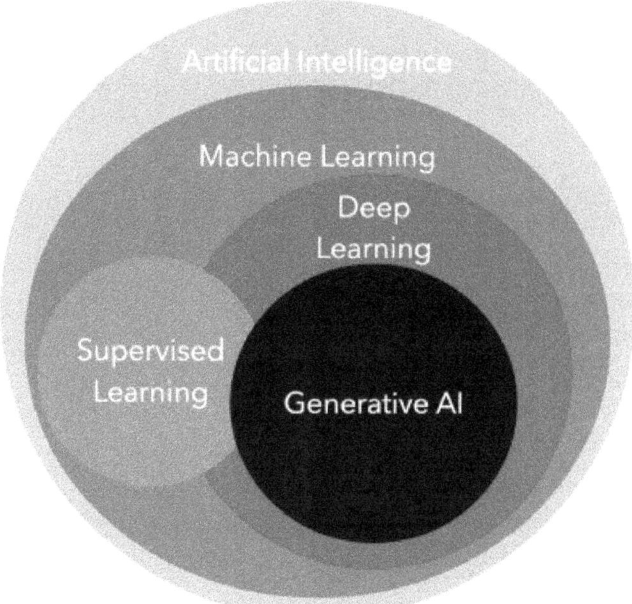

Figure 4.2: A taxonomy of GenAI-related disciplines.
Source: - (Singh, 2023)

<u>Importance and Applications of Generative AI</u>

- **Creative Content Generation:** Images, songs, poetry, and artwork may all be automatically generated using generative AI. It can aid artists, designers, and musicians in exploring new ideas, generating inspiration, and enhancing their creative processes.

- **Data Augmentation:** Machine learning makes use of generative AI to enrich sparse datasets with generated data. When getting massive volumes of actual data becomes a challenge, this helps machine learning models generalise and perform better.

- **Image and Video Editing:** Generative AI can assist in advanced image and video editing, allowing for tasks like inpainting missing regions, style transfer, and even converting images between different domains (e.g., turning a daytime scene into a night-time one).

- **Anomaly Detection:** Generative AI is helpful in quality assurance and control, forgeries, and security because it absorbs the characteristics of usual data, which makes it easier for it to detect unusual features in a set of data.

- **Image Synthesis and Style Transfer:** The generative AI models including GANs have greatly enhanced this field of picture synthesis. They are able to generate photographs of characters that are realistically indistinguishable from real objects, places or even faces of real people. Computer generated imagery, virtual reality and generating contents such as games and movies are all area of applicability for this technology. Another great use case on some of the generative models is style transfer where one can apply the style of a painter to a picture. While it is popular nowadays mostly in the sphere of creative photo manipulation and art.

- **Text Generation and Natural Language Processing:** The text creation and NLP capabilities of generative AI have been really astonishing. Language models based on big piles of literature including as OpenAI's GPT variants are able to generate human-like prose, poetry, and even the natural-sounding response to questions.

Challenges and Limitations

Today, generative AI has advanced a lot and explored certain opportunities, but it encounters some problems and drawbacks. Here are some of the key issues:

I. Training Instability and Mode Collapse

Some of the common issues with generative models especially GANs include training instability. Training GANs may be difficult because the process is sensitive to hyper parameters and network architectures. TWhen GANs are not trained further, they possibly suffer from mode collapse, a situation where the generator is not able to provide adequate variance since it has limited numbers of output. To overcome these problems many advanced works are proposed for various kinds of modifications in loss functions, architectures of GANs, and regularization methods for providing better and stable training of GANs.

II. Evaluation Metrics for Generative Models

Instead, the performance of generative models has been cumbersome to assess due to the lack of a ground truth value. Unfortunately, the quality of the produced data cannot be assessed with traditional quality indicators including accuracy and precision-recall. The quality and the diversity of the produced images can be assessed, for example, by the likelihood of getting a photo with a specific subject, called the Inception score or by the Fréchet Inception Distance (FID). However, these metrics have their drawback and are not sufficient enough to evaluate generative models. Preliminary and reliable assessment indicators are still an object of further explorations.

III. Ethical Implications and Potential Misuse

It is also unethical to use generative AI and particularly GANs to generate deep fake contents that look so real. To be more precise, one can use deep fakes for disinformation, constructing a news story or tarnishing someone's reputation. The ethical questions and the regulation of the use of the generative AI have to be solved and rules and laws for not using the negative sides of the technology have to be created as it develops. AI that is responsible in creation and use helps differentiate between generative AI that is useful and that which may cause harm to society.

IV. Data Privacy and Security Concerns

Large generative AI models are also capable of repeating information that it was trained on, and thus it may memorise sensitive information. Both privacy and security are at risk since these models could reveal private information about persons or organisations.

Differential privacy and data anonymisation reduce these dangers. Preserving data confidentiality and guaranteeing that generative models are not damaging any sensitive data becomes an issue; more so given that generative AI models are becoming sophisticated in detecting data patterns.

If it is about Responsible research, development and implementation of generative AI then there must be solutions to these constraints and difficulties. To get the best out of generative AI as well as to eliminate any dangers that come with it, they include the following issues where researchers, lawmakers, and stakeholders have to address as the technology grows (Kumar & Sharma, 2024).

Robotics and Automation:

Robotics is the study of robots, which are machines that are capable of automatically performing a series of complex actions. The study of robots, including their creation, programming, and use, is known as robotics. Conversely, automation refers to the process of using technology to carry out activities automatically, without the need for human involvement. A wide range of technologies, including robotics, AI, and ML, may be used in automation processes.

The field of robotics within AI is dedicated to the study of machine intelligence and its development. Due of its mechanical build, electrical components, and programming language, robotics brings together electrical engineering, mechanical engineering, and computer science and engineering. Though the two fields aim to accomplish distinct things and have different uses, the general public often considers robotics to be a subset of AI. If given artificial intelligence, robots may mimic human behaviour and even seem like people.

AI and robotics are connected via artificially intelligent robots. ML, computer vision, RL learning, and other AI technologies are used by AI programs to drive AI robots. The majority of robots are often not AI robots; instead, they are designed to carry out repeated motions and don't need AI to do their tasks. Nevertheless, the capabilities of these robots are restricted.

Applications of robotics and automation

1. **Manufacturing:** Robots are widely used in the manufacturing industry to assemble, package, and inspect products. They are also used for material handling, welding, and painting. Technology is applied in the production process by minimizing on mistakes that

may occur, enhance Timely production, and cutting down the general cost.

2. **Healthcare**: Robots are applied to the health care field to perform operations, treat patients and help people with disabilities. They are also employed in the support role in patient care and in such activities like distribution of Medicines.

3. **Agriculture**: Application of robotics and automation are in operations like planting, harvesting and use of pesticides in farming. This is done in an effort to cut costs and particularly the cost of human resource in organizations.

4. **Transportation**: Self-driving automobile and trucks are appearing on the roads. They are created for improving safety and minimizing the role of the human factor in traffic.

5. **Education**: Control technologies known as Robotics and automation are also applied in education to train students in areas such as programming, engineering and Robotics.

Impact of robotics and automation

1. **Greater efficiency**: Robots and automation can work faster and more accurately than humans, leading to increased efficiency and productivity.

2. **Cost savings**: Utility can reduce various costs and especially labour cost since automation reduces the demand for human employees. For companies, this implies the possibility of cost reduction.

3. **Improved safety**: Robots can also do dangerous operations which humans cannot, and thus insuring everyone's safety. Also, as with any process automated there could be fewer chances of errors due to human error.

4. **Job Displacement**: Though robots and automation can lead to productivity improvements they also have the negative effect of job elimination.

5. **Skills required**: This means that the skills required to plan, implement, control and maintain robots and automation equipment will gain more demand as these technologies advance.

1. AI in 5G Networks:

Artificial Intelligence (AI) plays a significant role in addressing issues that conventional methods cannot handle and that need a large number of factors that are inappropriate for hard-coded software to set up. By identifying certain patterns that a person would miss, AI approaches can also learn from and adapt to changing situations. Despite their apparent lack of relevance, the phrases AI and the 5G technological standard must be taken into consideration within the same framework. AI methods might efficiently meet the high bandwidth and low latency needs of 5G systems while managing the heavy network traffic. For current 5G systems to satisfy performance and reliability requirements, they must be appropriately optimised. Mobile and wireless communications challenges that are neither linear or polynomial are many, and solving them calls for a variety of AI approaches. stress that AI can fulfil the cellular networks' 5G standardisation standards.

Thus, autonomous driving and remote surgery are two areas that may benefit from URLLC service. The fourth service, known as mMTC, is all about systems that have a lot of tiny, concentrated devices in one place and use very little bandwidth. The key features of the 5G categories are shown in Figure 4.3.

5G key scenarios	Key system performance requirements	Traffic features and challenges
eMBB	Peak data rate: DL: 20 Gb/s, UL: 10 Gb/s User experienced data rate: DL: 100 Mb/s, UL: 50 Mb/s (dense urban) Area traffic capacity: DL: 10 Mb/s/m^2 Latency: 4 ms	High data rate every user; downlink-dominated transmissions traffic; big data package transmission
URLLC	Latency: 1 ms Reliability: 99.999 percent	Strict requirements on latency and reliability; frequently short path, small package transmission
mMTC	Connection density: 1 million deveices/km^2	Low data rate every user; high connection density; uplink-dominated transmissions

Figure 4.3: Characteristics of different 5G services

Source: - (Glisic & Lorenzo, 2022

2. Self-Healing Networks:

With its ability to autonomously identify, diagnose, and recover from network defects without human involvement, self-healing networks constitute a paradigm change in the area of communication systems (figure 4.4). The proactive detection and mitigation of disturbances is built into these networks, guaranteeing uninterrupted operation and peak performance. There are several important parts to the idea of self-healing networks, and they all work together to handle faults. As innovative as it is revolutionary, techniques in self-healing networks allow for the detection, diagnosis and treatment of network faults without any need for intervention from professionals. These networks are designed in advance with special intention to minimize the interruption to zero level and to maximize the networks performance. Self-healing network in fault management encompasses many critical elements as follows. In the world of communication systems, self-healing networks can locate, analyze, and cure current network malfunctions on their own, without the input of a human operator. These networks are intended to identify disruptions before they occur so that it can continue to operate at optimal efficiency uninterrupted. The idea of self-healing networks includes several important parts, all of which are vital to the fault management procedure. Originally, when a physical or logical failure occurred in a traditional corporate network, the design would immediately switch over to standby gear and any related backup connections. A major drawback of these redundancy methods was their lack of insight into the specific traffic flows that were impacted. Compared to conventional networks, self-healing networks provide a great deal more advanced redundancy feature. When there is a catastrophic loss of network connection, these networks depend on AI, machine learning (ML), and automation to keep traffic running. Network service deterioration may be recognised or even expected by self-healing networks, which aspire to optimise, forecast, and automatically act. The advent of self-healing networks has been a game-changer in the realm of communication systems as they can identify, fix, and recover from network failures independently of humans. These networks are meant to identify the opportunities of an interruption before it occurs, so that they can continue doing what they do best uninterrupted. There are several sub-ideas within the self-healing net-works concept which all contribute to dealing with faults. Another concept for enhanced system availability and to compensate for the impact of network faults is that of self-healing networks. A critical characteristic of self-healing networks involves decentralization of anomaly detection and fault diagnosis as well assembled implementation of remedial actions. The use of advanced technologies particularly in artificial intelligence is vital to self-healing networks. This means that AI provides ability for detection and rectification of faults to these systems. Thousands of models and algorithms help AI-based solutions analyze lots of network data and, recognizing signs of potential issues, make rational decisions that help minimize disruptions. In the context of fault detection and subsequent self-healing mechanisms, this work explores how AI-based approaches are useful to auto-healing networks.

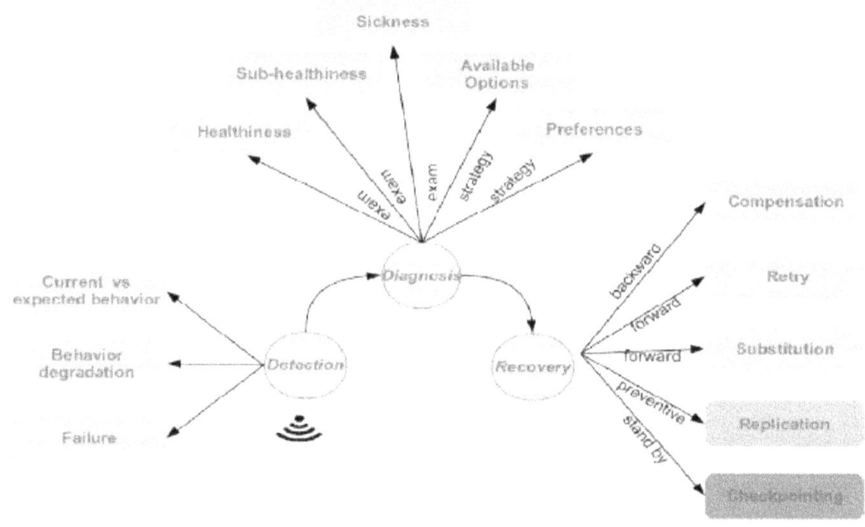

Figure 4.4: Self- Healing Network Strategy
Source: - (Nand Kumar, 2023)

3. Federated Learning (FL).

Collaborative learning, or FL, is a machine learning approach that builds an algorithm across many decentralized edge devices and delivers AI models to the data source. Only the updated model-related data may be sent in this kind of learning; the training data are not transferred across different sections. With this method, consumers will have the benefit of having access to a large amount of data without the need for central storage. The aforementioned characteristics are shown by a multitude of apps that integrate intelligent features in the mobile arena, including voice recognition, language modelling, and picture classification. In fact, even while sending anonymized data to the central data repository, users are still vulnerable to dangers. In contrast, FL successfully mitigates these risks by communicating just the information that is necessary for improving the model. The FL framework using the federated averaging (FedAvg) procedure is shown in Figure 4.5. The FedAvg process at a central FL server, FL communication, and local FL device processing are the primary components of the FL process.

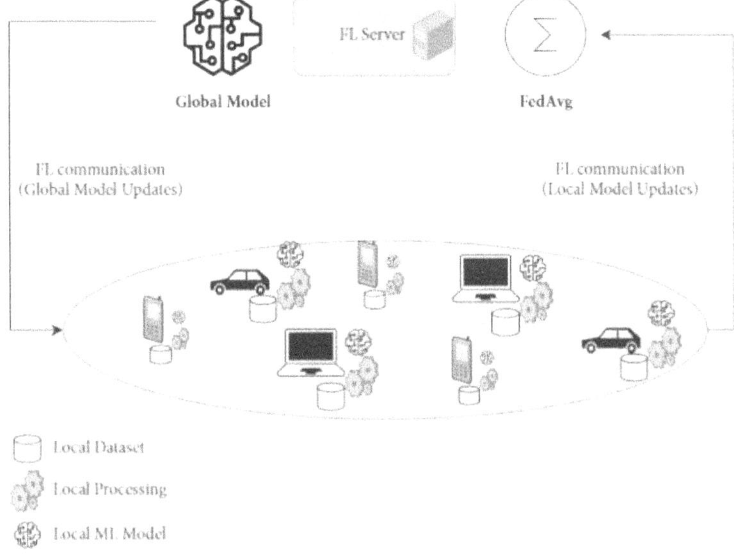

Figure 4.5: FL framework with federated averaging

Source: - (Akhtar et al., 2020)

4.2 Edge Computing and IoT Integration

A new age of computing architecture is being ushered in by the combination of Edge Computing with the IoT, where data processing and analysis take place closer to the data source. The conceptual framework of edge computing inside the IoT is examined in this part, along with the benefits of this integration and instances of applications that profit from this mutually beneficial interaction. Distributed computing is at the heart of Edge Computing and the IoT. In this setup, computing resources are placed strategically at the edge of the network to enable localized data processing and analysis. IoT devices gather data and send it to neighboring computer nodes or edge devices in this architecture. These edge devices on their own based on certain programmed rules or machine learning algorithms can respond to data, process, filter and analyze it for useful insights in real time. Data processing may result in its migration to centralized cloud servers for storage, for future use or to other business systems. The rise of edge computing as a paradigm shift in contemporary computing has been driven by the need to tackle the problems caused by the exponential growth of the internet of things (IoT) and the exponential increase in data generated. Research into edge computing frameworks and architectures is being propelled by the pressing need to better process data, decrease latency, strengthen security, and effectively handle the massive amounts of data generated by the IoT and big data. Managing the vast amounts of data generated by IoT devices is usually accomplished via the architecture of the cloud. Latency, higher bandwidth needs, and long transmission times are just a few of the problems that cloud computing faces. An emerging solution to these problems is the idea of edge computing. As such, this method enhances scalability, latency, and privacy because it enables the handling of data without subsequent transmission to a central server.

1. **Addressing Latency and Real-time Processing:** Faster and more efficient processing helps to learn some applications as such as smart city, driverless cars, or industrial automation. To do this, edge computing servers perform data processing near the source as this enables faster decision making. Researchers Shi et al. noted in

their study that edge computing enables low-latency interactions in IoT applications by reducing communication latency.

2. **Bandwidth Optimization:** Data transmission efficiency and network capacity are threatened by the enormous deluge of data produced by IoT devices. Due to edge computing, less data have to be sent to the server as the processing is done locally. In their extensive examination of edge computing, they go into detail about this bandwidth optimisation.

3. **Privacy and Security Enhancements:** By handling sensitive data locally and minimising exposure during transmission, edge computing provides a solution to the growing worries about data privacy and security. delves into the ways in which edge computing might improve the security and privacy of data in IoT projects.

4. **Resource-efficient Data Processing:** A distributed processing strategy is typically required due to the limited computing capacity of IoT devices. In order to maximise efficiency, edge computing moves computation to nodes at the network's periphery. Touches on the topic of making the most of computing resources in edge situations.

5. **Offline Capabilities:** Applications in outlying locations or mission-critical situations greatly benefit from edge computing's offline functionality. The importance of offline capabilities in edge computing is brought to light by the study undertaken by Bormann and Castellani within the framework of the Internet Engineering Task Force (IETF).

6. **Scalability and Flexibility:** Computing solutions must be scalable and agile to accommodate the ever-changing nature of IoT settings. According to Bonomi et al., researchers may learn about edge computing architectures and use that information to create systems that can scale up or down depending on local demand without sacrificing performance.

7. **Seamless Integration with Existing Systems:** It is critical to comprehend the difficulties of interoperability when edge computing incorporates with pre-existing cloud-based

infrastructure. gives information on how to combine cloud computing with edge computing.

Architecture of Edge Computing-Based IoT

The main goal of edge computing within the framework of the IoT is to reduce network traffic and decision-making latency in a variety of IoT situations. The IoT architecture that relies on edge computing has three separate levels: the IoT, the edge, and the cloud. These layers are constructed using pre-existing edge computing reference designs. Determining the precise roles played by each layer and investigating the established means of communication between them are our primary concerns.

Consists of three distinct layers:

1. **IoT layer:** A wide variety of physical items are included in the IoT layer, including smart vehicles, robots, smart machinery, portable terminals, instruments, meters, and other similar items. The operation of services, activities, or equipment is overseen by these things. In addition, the management of computing resources inside IoT devices is made possible by the actuators, sensors, controllers, and gateways composed of the IoT layer, which is specifically designed for IoT settings.

2. **Edge Layer:** In addition to offering real-time services like cognitive computing, security, and analysis, the Edge Layer forms the foundation for receiving, processing, and forwarding data streams that originate from the Device Layer. It comprises three sub-layers based on processing capability:

    1. **Far-Edge Layer (Edge Controller Layer):** This layer gathers IoT data from edge controllers, completes initial threshold checks or filtering or passes control of the device flow back to the IoT layer. An edge controller is integrated with application-specific algorithm lookup tables and filters data prior to decision-making and relays decisions to connected IoT devices through PLCs.

2. **Mid-Edge Layer (Edge Gateway Layer):** Including edge gateways which link to the wired and/or wireless networks this layer takes data from the edge controller layer, caches it and hand over control between the layers. With higher capacity, it provides comprehensive IoT data processing and tracking devices across the multi-layer.

3. **Near-Edge Layer (Edge Server Layer):** There, advanced data processing is conducted using highly potent edge servers. These servers are data collectors from edge gateways, decision-making entity, and platform management and business application hosting.

3. **Cloud layer:** Extensive data mining is the main emphasis of this layer, which aims to optimize resource allocation on a national, regional, or even organisational level. By way of the public network, data is sent from the edge layer to the cloud layer. Companies' business apps, services, and model deployments hosted in the cloud may also provide feedback to the edge layer (Enoch Oluwademilade Sodiya et al., 2024)

Key benefits of edge computing in IOT

1. **Real-time Latency Reduction:** In essence, many IoT apps are sophisticated monitoring systems; they gather data, process it, and then act on the insights they provide. others devices do this automatically every hour, others do it every day, and yet others do it only when prompted by a user action. In situations when these insights are required immediately, edge computing may be useful for the IoT. Data collection and analytics are done at a physically closer location, often within the same nation or area, or even on the premises, instead of in a huge centralised data centre, by bringing computation closer to the IoT device. Due to the shortened round-trip time to the data centre and return, network latency is minimized this way.

2. **Enhancing IoT Security:** Managing security in an environment where more and more devices are linked is a big concern for many

in the IoT ecosystem. One-way malware may exploit IoT devices is to launch DDoS attacks. Although the security of edge computing cannot be guaranteed to be higher than that of a private cloud, the advantages of being closer to users are undeniable. Companies who are worried about data storage in places with differing data protection rules than data generation locations may find some relief with edge computing.

Current State of Edge Computing in IoT

Edge computing is now experiencing a period of explosive growth, broad acceptance, and constant innovation as part of the IoT ecosystem. This part gives a survey of the current state of affairs, describes the obstacles encountered by Edge Computing in IoT settings, and identifies adoption patterns and business applications.

There is a wide variety of technologies, platforms, and solutions that make up the current Edge Computing landscape in the IoT. Their goal is to facilitate decentralized data processing and analysis. Businesses in a wide range of sectors are beginning to see how Edge Computing may solve problems with bandwidth, latency, and privacy that have long plagued cloud-centric designs. This has resulted in growth of Edge Computing facilities in some sectors such as smart city, health, transportation and energy sectors. Technology behemoths, upstarts, system integrators, and telecoms all play an important role in the Edge Computing arena, and they all provide solutions that are unique to their respective use cases and industries.

The expanding number of IoT devices, the increasing need to analyze data in real-time, and the need for edge-based intelligence are all drivers propelling the use of edge computing in the IoT. A wide variety of use cases reported in industries that edge computing applications enhance the monitoring and control of manufacturing processes in real-time, maintenance and quality control, and enhance operation performances with reduced or eliminated down time." Edge computing enables combination of medical devices, remote patients monitoring, tele medicine in health care delivery improving the health status of patients by delivering personalized treatments." It empowers city planners with optimum infrastructure solutions and effective decision making because of various advanced smart city applications based on edge computing plans. Such activities involve surveillance in areas of security, order and environment for instance traffic control.

Consumers may benefit from Edge Computing in terms of happiness and revenues realized through appealing personalization of suggestions, stock control and distributions. Improved road safety and reduced congestion are two outcomes of edge computing's assistance for applications including vehicle-to-vehicle communication, autonomous driving, and traffic optimisation.

There are also several challenges associated with edge computing in IoT settings even though this model is more imminent and realistic. One of these is the realization that mostly edge devices have limited processing capabilities, local memory and storage to execute complex analytics and ML algorithms. Other security challenges in edge computing includes Data exposure, Device tampering, unauthorized access that calls for high security measures to enhance the security of sensitive data besides reducing risks of cyber threats. Concurrently and concurrently connecting various devices, protocols, and standards in an Edge Computing environment is challenging, which hinders flexibility and scalability (Enoch Oluwademilade Sodiya et al., 2024).

4.3 Blockchain in Secure Networking

Technology that deals with deals that include virtual and physical commodities and contracts under a secure and traceable system is referred to as a blockchain. When it comes to protecting networks, blockchain technology is both effective and inexpensive. Integrating many fundamental technologies including cryptographic hashes, digital signatures, and distributed consensus methods, it functions as a decentralized ledger to record all committed transactions in trustless contexts. A wide range of network interaction systems, including public services, social networks, reputation systems, smart contracts, the IoT, and financial and security services, have made use of blockchain technology in recent years. There has been a surge in interest in exploring the potential of blockchain technology to resolve network security issues and vulnerabilities and to comprehend the practical security consequences of this technology's broad use. Financial transactions, medical information, property titles, and any other valuable asset may be recorded and tracked using a suite of distributed ledger technology known as blockchain. The distributed ledger system, which blockchain builds upon, has been around for a long time. Simplified, it is a digital ledger that records all kinds of transactions that take place in a network where nodes are directly connected to one another. The general consensus is that this technology will "cut out the middleman" whenever digital assets are traded or purchased. Because it is decentralized, this medium is much safer. Institutions in the financial sector are investigating the feasibility of using this technology to guarantee the safety of financial transactions.

The distributed ledger technology known as blockchain has several features with traditional digital ledgers, such as:

- Blocks can only be appended;
- No block can be edited;
- To prevent fraud, the legitimacy of each transaction is dependent on the one before it;
- Transaction happens only after full verification.

Blockchain may be summed up as a public ledger that contains a list of blocks that include various components of a transaction. Along with the chain's immutability and resistance to manipulation, its size grows as new blocks are added. Examine how blockchain networks function to have a deeper understanding of how blockchains function.

This network uses copies that are sent to each node to depict a group of nodes as belonging to the same blockchain. These nodes constitute a peer-to-peer network where:

1. A private and public key pair allows every user to communicate with the network. Both user authentication and transaction signing rely on such keys. All of the nodes are informed of each signed transaction. To ensure the network's integrity, authenticity, and non-repudiation, blockchains use symmetric cryptography.

2. The distribution over the whole network occurs when each neighboring node validates the transaction before sending it on to the next node. We discard invalid transactions.

3. A candidate block is created by packaging validated transactions and adding a timestamp. "Mining" describes this procedure. After then, it's the mining node's job to broadcast the network this timestamped block.

4. After the block's validity as a repository for transactions is confirmed, a hash referencing the preceding block is appended to the chain, and the node's environment is refreshed.

Figure 4.6 depicts the aforementioned procedure, which is the adding of a verified block to the chain.

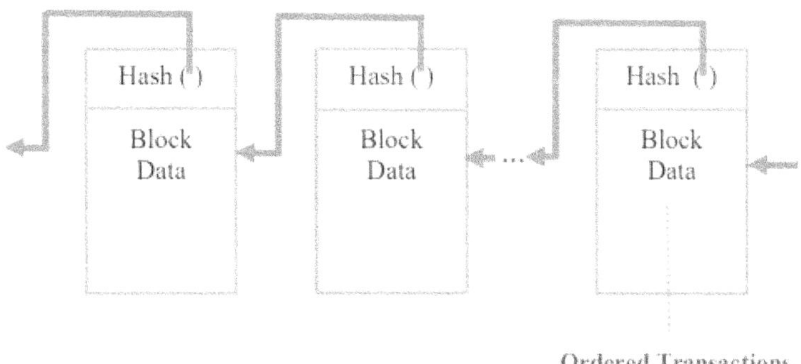

Figure 4.6: The Blockchain Structure.

Source: - (Liu et al., 2021)

<u>Characteristics of Blockchain</u>

The most well-known and effective application that raised awareness of blockchain technology is Bitcoin. But as time has gone on, the IT industry has come to understand blockchain's full potential outside of the banking and bitcoin industries. Therefore, this section discusses the special qualities and attributes of blockchain technology that set it apart from the current database management system.

I. Immutability

It is impossible to change or manipulate a transaction once it is uploaded to a blockchain network. This distinctive feature of blockchain eradicates the prevalent misunderstandings around the preservation of digital records, including the possibility of data alteration, deletion, or tampering. Therefore, a new block with the proper facts must be inserted in the event that a transaction is entered with mistakes; it cannot be erased or rewritten. The participants would also be able to see the initial mistaken transaction alongside the newly added one, with the correct timestamp shown in chronological order.

II. Decentralization

Blockchain depends on a network of nodes rather than a single controlling authority or central server to handle or store data because of its decentralisation feature. Consequently, in a blockchain network, each node serves as a server that may store crucial digital data that can be accessed with a private key, such as cryptocurrency, financial transactions, contracts, or anything else.

III. Distributed Ledger

Distributed ledgers allow all nodes in a network to access and update the same database at the same time. In contrast to the centralised nature of conventional banking and financial activities, blockchain technology is based on a distributed network of computers called nodes. Transparent and trustworthy data is created when every node in the network has a copy of the transaction and it takes a majority vote to add additional transaction information. One further benefit of a shared ledger is that it prevents duplicate entries since each transaction may only be recorded once. Not every distributed ledger is blockchain, even if blockchain is a kind of distributed ledger.

IV. Consensus Mechanism

Blockchain technology and other distributed or multi-agent systems rely on a consensus mechanism. Since blockchain does not rely on any one entity to confirm or validate transactions, it relies on a consensus method whereby the network nodes agree to approve or reject the transaction. Consensus, in its simplest form, is the process by which all of the participating nodes in a blockchain network agree to verify a transaction. This is crucial for fostering confidence in a distributed system and guarantees that, particularly in a public blockchain, only authentic data is added as a new block. When the consensus mechanism becomes more complicated, however, the rate at which new blocks may be added decreases. Therefore, a private or permissioned blockchain with a comparatively easier consensus method may be employed for a big industrial application like a supply chain since the right to add new blocks is with fewer known parties. In the event that there are several failed nodes, the consensus mechanism also aids in maintaining data consistency. There are two different kinds of failure nodes: Byzantine and crash fault nodes. Byzantine fault tolerance ensures the network can continue to operate even while hostile nodes are present, whereas crash fault tolerance ensures that consensus can be reached even if certain components fail. Blockchain is called a "thrustless system" because, unlike a centralised agency, the participants of the network execute transactions based on consensus achieved between them, even though trust is built on data based on distributed ledger technology. This is particularly true in a public blockchain where the participants do not know each other. This is a concise overview of the several consensus algorithms employed by blockchain.

<u>Core Components of Blockchain</u>

Decentralization: Blockchain relies on decentralization rather than centralization, with each participant (hub) storing a full copy of the blockchain. You won't need a middleman in a decentralized network to have direct conversations with your friend. That was the fundamental principle behind Bitcoins. Your financial situation is entirely within your control.

Immutability: Data that has been added to a blockchain cannot be removed or altered once added. Tools for cryptographic hashing and agreement provide this persistence. In the context of blockchain technology, immutability refers to the fact that data cannot be altered once added to the blockchain. The cryptographic hash function is responsible for the blockchain acquiring this attribute. Hashing, in its most basic definition, is the process of producing an output string of a predetermined length from an input string of any length. **Transparency:** Transparency and accountability are guaranteed by the fact that all contributors may see all conversations on a blockchain. A term that is both fascinating and sometimes misunderstood in blockchain is "transparency." Complex encryption ensures that only a person's public address may be used to represent them.

Consensus mechanisms: Several protocols secure and encrypt communications between nodes (e.g., Proof of Work and Proof of Stake).

The Role of Blockchain in Secure Networking

Blockchain addresses key challenges in networking security and offers transformative solutions:

Why blockchain for networking security?

- **Decentralized Trust Model**: Here, blockchain displaces centralized power structures by decentralizing trust over a network of nodes. Every node is involved in the processes of reaching consensus to transactions, thus eliminating concentration.
- **Tamper-Proof Data Storage:** Another important attribute of the Blockchain is that none of the data entered into the system can be edited or deleted. This also prevents unauthorized modification while also ensuring that data is not modified in any way.
- **Enhanced Transparency:** Each transaction is made and stored in a public ledger to which only approved parties have access to. This

transparency enables accountability and will facilitate traceability of network operations.

Blockchain-Based Network Architecture:

- **Node Communication:** Each node in a blockchain network safely communicates with other nodes using cryptographic means. It also means that there is privacy and the information being received is genuine.
- **Consensus Mechanisms:** Other approaches such as Proof of Work (PoW) and Proof of Stake (PoS) legitimize transactions for reliable running of the system.
- **Decentralized Control:** Again, there is no central, or focal, controlling or regulating position that would make the network vulnerable.

Advantages of Integrating Blockchain into Networking

- **Increased Security:** Blockchain includes cryptography techniques to protect the information and the flowing messages from penetration.
- **Operational Resilience:** The decentralized nature of the runtime means that the system will work even if a number of nodes don't.
- **Cost Efficiency:** Blockchain removes the need for middlemen and increases reliability, which decreases operational expenses due to a lower likelihood of cyber assaults.
- **Real-Time Verification:** It provides confirmation in real-time for all the transactions and exchanges of data which makes the process faster and secure.

Opportunities and Challenges

Blockchain technology offers great promise in any service that deals with data or data transactions. Cyber physical systems, healthcare industries, end-to-end messaging services, cloud data services, e-commerce, mobile commerce, banking, the internet of things, data analytics, healthcare infrastructures, and medical data exchange are all included.

The aforementioned domains are not the only ones where blockchain technology has been used. In addition, blockchain technology generates a number of new job opportunities in several industries, including:

- **Blockchain Developer** – The major tasks are planning, executing, and providing ongoing support for a blockchain-based network as it goes through its development lifecycle. Building and launching a blockchain network, designing blockchain technology around a specific business model, researching new technical solutions and protocols, automating blockchain development workflows, implementing test-driven development practices, and analyzing requirements are all additional tasks that will need to be completed.

- **Blockchain Engineer** – The primary responsibilities of a blockchain engineer include designing and implementing the network's backend, constructing products that run on top of it, building meta-transaction infrastructure through mobile apps, creating frameworks and infrastructure for digitally signing documents, and creating APIs that connect to the blockchain.

- **Blockchain Platform Engineer** – The blockchain platform engineer's responsibilities include developing knowledge of blockchain network designs, platforms, and administrative and operational needs, as well as providing development assistance and expertise to blockchain projects throughout the organisation. He/she designs and builds

Applications of Blockchain in Networking

1. Decentralized Identity Management

Conventional identity management entails the use of databases that are normally protected in central locations and hence vulnerable to attacks. Through decentralization of identity, blockchain allows users to own their identities through self-sovereign identity systems. Due to blockchain's ability to keep records immutable, these systems use this to authenticate identities without intermediaries.

2. Securing IoT Networks

The Internet of Things (IoT) – billions of interconnected devices introduce immeasurable risks:

- Devices should first be authenticated before being allowed to the network.
- Capture data on device transactions on the blockchain accurately and securely.
- Protect IoT firmware from being tampered. nitty management relies on centralized databases prone to breaches

3. Protecting Domain Name Systems (DNS)

DNS, often targeted by cyberattacks, can be secured using blockchain's distributed ledger. By decentralizing DNS records, blockchain ensures:

- Resistance to DNS spoofing.
- Enhanced reliability and availability.

4. Mitigating DDoS Attacks

Blockchain's decentralized architecture can absorb and mitigate DdoS attacks by distributing traffic across multiple nodes. Tokenized mechanisms also enable fair resource allocation.

5. Securing Data Transmission

Blockchain ensures that all data transmitted over the network is encrypted and recorded immutably. This creates an auditable trail for sensitive communication.

4.4. Future Innovations on the Horizon

Quantum computing & 6G technologies can be expected as disruptive technologies that would reshape the new era AI Networking & communication. Quantum computing provides unmatched computational advantages which can act as a breakthrough in data analysis, encryption and networking. At the same time, the integration of AI with the upcoming 6G offers super-high bandwidth, strictly minimal latency, and the possibility of intelligent edge computing that can enable applications such as real-time holo-telepresence and the subsequent generation of the IoT. In their capacity, these technologies belong to an emerging era of smarter, faster and secured computer networks.

4.4.1 Quantum Computing's Role in AI Networking

Quantum computing holds the potential to revolutionize AI networking by introducing unparalleled computational capabilities that could significantly transform how data is processed, secured, and optimized within networks. In the 1980s, the fields of quantum mechanics and information theory were brought together, leading to the development of quantum information theory. This was the theoretical genesis of quantum computing. New fields including quantum computing, communication, and cryptography emerged from the application of quantum mechanical concepts to the modelling and processing of information. This provides a synopsis of relevant scientific case studies as well as an introduction to the aforementioned ideas. The potential for certain quantum phenomena to significantly outperform conventional computing speeds was a driving force behind the creation of quantum computing. The fundamental building block of quantum computation, the qubit, can take on a continuum of values, distinguishing it from conventional computing and enabling the storing and processing of quantum superpositions of data. This technological synergy is likely to enhance the following dimensions:

1. **Data Processing Speeds:** The developments that inexorable are, such as Shor's and Grover's quantum algorithms, make it possible to work with very large sets of data at high speeds. This ability ensures that data are analyzed and decisions are made on time particularly for networks supported by artificial intelligence that changes frequently and demands prompt response.

2. **Cryptographic Security:** Hiding algorithms, which are lattice- and multivariate polynomial-based, provide ways of avoiding yielding to threats from quantum decryption operations. These

methods ensure that any information exchanged via connected AI networks remains safe despite the ongoing improvement in the quantum computing proficiency.

3. **Network Optimization:** This technology can address complex optimization issues that few other technologies are capable of today. These include issues related to network architecture, traffic control and resource availability which creates an opportunity for the development of intelligent AI self-organizing networks.

However, there is evidently some key challenge when it comes to apply quantum computing in AI networking. There are challenges such as high implementation costs, the absence of quantum-safe algorithm standardization and the present weak links in quantum hardware infrastructure working against the full adoption of quantum computing.

4.4.2 The Convergence of 6G and AI

The emergence of 6G technology expected to offer nearly zero latency, extremely large bandwidth, and connectivity practically everywhere will be grounded on AI. The convergence of 6G and AI will redefine network operations and enable ground breaking applications. Key areas of impact include:

1. **Autonomous Network Management:** 6G would implement AI autonomic computing techniques that would enable network components to learn, optimize and adapt their resources and performance over time to support the various requests made to the network as well as minimize energy consumption.

2. **Predictive Analytics:** With AI system incorporated into 6G networks, they can learn from previous occurrences of network problems like congestion or failure and other such problems in order to prevent their occurrence.

3. **Edge Intelligence:** Also, at the edge of the network, with the help of AI, the intelligence in the 6G can be distributed and thus the latency of data transfers will be less, making real-time decisions

possible for applications such as autonomous cars, smart cities and industrial applications.

This convergence is ready to open the floodgate of revolutionary applications that would comprise real-time holographic communication, enhanced IoT systems, and Ary applications. However, achieving these advancements will require:

1. **Advanced AI Models**: Sophisticated algorithms capable of handling the immense complexity of 6G operations.
2. **Robust Blockchain Frameworks:** Security frameworks leveraging blockchain for safeguarding data integrity and preventing cyber threats.
3. **Collaborative Standards Development:** Industry-wide efforts to establish interoperability and infrastructure that aligns with global standards.

The use of AI with 6G requires interdisciplinary knowledge and cooperation, policy initiative and strategic planning with strong emphasis on ethical and sustainable implementation of technologies.

<u>Challenges and Future Trends in 6GAI-Enabled Wireless Communications</u>

The employ of 6G technologies and artificial intelligence (AI) brings in a large number of opportunities for the future wireless system but also it creates a lot many issues. Looking at what 6G offers; it will be a world of ultra-fast and intelligent connectivity under the leadership of AI automation and optimization, 6G faces various challenges which include data privacy and energy requirements for networks, and the overall challenge of handling decentralized network. Examining these challenges hand in hand with new trends will define the direction toward the achievement of the optimum utilization of the 6G-AI in wireless communication which will open up potential opportunities across dynamic industries.

Challenges and Complexities:

1. **Massive Data Handling.** Efficient management of storage, processing, and bandwidth needs will be crucial for 6G networks to create and handle large amounts of information, ensuring continued competitiveness in today's globalized society. More and more gadgets are contributing data, which might cause a revolution in data processing with 6G mobile technology. The use of AI in fux data management is crucial [7, 163, 164]. 6G networks can safely and rapidly transport massive amounts of data produced by IoT devices and vehicles. Edge computing's local processing support further optimises latency management, and AI storage solutions play a crucial role in prioritizing and categorizing this data to decrease latency. Analytics powered by AI sift through this mountain of data in quest of insights that let decision-making in real-time, such as optimizing traffic or allocating resources.

2. **AI Algorithm Complexity**. Complex artificial intelligence algorithms provide an impossible barrier to real-time implementation on 6G networks due to the enormous processing power and hardware efficiency requirements. Developing and deploying sophisticated AI algorithms over these networks flawlessly in real-time is an absolute need. As 6G networks grow to manage enormous volumes of data, more complex AI algorithms will emerge, necessitating optimised hardware and huge processing capacity for AI development. The development of 6G networks will allow them to more easily manage such loads. To ensure flawless real-time performance, these algorithms—which include DL-NN and RL models—need a lot of processing power and need to be optimised. This objective can only be achieved if the hardware is optimised for these algorithms to provide the best possible real-time performance. Efficient hardware is essential for the success of AI computations, as they are highly compute-intensive. To tackle complex calculations effectively, we may need

high-performance GPUs, specialised AI accelerators, or maybe even quantum computing.

3. **Privacy and Security**. 6G landscapes are greatly influenced by AI and ML, which give it the ability to learn from dynamic, ever-changing environments. But like two sides of a double-edged blade, this AI and 6G partnership has both benefits and downsides. AI technology has the potential to enhance 6G's privacy and security features in a number of ways, but it also presents concerns associated with security lapses that seriously jeopardise the technology's viability [171]. Concerns about data privacy, security flaws, and the moral implications of AI-generated insights are legitimate as AI-enabled networks usher in a new age of connection. Additionally, worries about AI/ML-powered 6G applications continue [28]. Data security, preventing unauthorized access, and reducing the likelihood of vulnerabilities affecting AI-powered systems in 5G and beyond are all important issues. International Journal of Intelligent Systems Despite the revolutionary benefits that these networks may provide, the hazards associated with them must be thoroughly evaluated in order to provide a safe and responsible online environment.

Future Trends

1. **Intelligent Spectrum Management.** 6G networks will heavily rely on intelligent spectrum management, which will completely revamp the allocation and utilisation of spectrum resources via the use of AI-powered services. AI plays an increasingly important role in dynamically optimising network performance and minimising interference in light of increasing connection demand and an oversaturated RF environment. Efficiency and underutilisation have resulted from the rigid nature of spectrum allocation in the past. AI has revolutionised networks by allowing them to react in real-time to changes in demand and interference patterns. Algorithms powered by AI can sift through mountains of data on

user actions, device kinds, and environmental factors to provide well-informed recommendations about spectrum distribution.

2. **AI-Driven Beamforming and Antenna Arrays**. Modern AI beamforming methods will solve the unique problems caused by millimetre-wave frequencies, expand coverage, and optimise signal delivery. As the 5G standards are nearing completion, investigations into potential 6G wireless network plans have started in earnest. To service providers, RISs are the most promising technology for the next 6G networks. These surfaces provide a system unprecedented degree of flexibility to form wireless channels, enabling it to tailor each channel's properties to its needs. To completely understand radiation pattern properties, however, a thorough understanding of how a meta surface acts in every potential operating scenario is necessary. Although both analytical models and extensive wave simulations require significant processing resources and have limits in certain circumstances, they may both be used to get a deeper understanding of radiation pattern attributes.

3. **Autonomous Networks.** 6G networks will become more autonomous with AI-driven self-optimizing, self-organising, and self-healing capabilities. The fast emergence of new apps and services as well as the ubiquitous usage of mobile devices have put tremendous demand on mobile networks. However, since networks are heterogeneous and becoming more so over time, managing these needs may be quite difficult. However, adopting cutting-edge network automation technologies, such as zero-touch management techniques, may be helpful in this situation. The integration of blockchain-based smart systems into a zero-touch pervasive AI as a service architecture in 6G networks can simplify its deployment across application and infrastructure domains, alleviating users of concerns regarding costs, security, or resource allocation requirements while satisfying 6G's demanding performance standards.

Multiple Choice Questions (MCQs)

1. **What is the primary role of Generative AI (GenAI)?**

   a. To generate new content similar to the training data
   b. To enhance the security of AI systems
   c. To improve machine learning algorithms' accuracy
   d. To automate the decision-making process

2. **Which of the following is a major challenge in computer vision that remains unsolved?**

   a. Creating 3D models from images
   b. Tracking individuals in real-time
   c. Interpreting an image at the level of a two-year-old, such as counting animals
   d. Recognizing faces accurately

3. **What is the key feature of Software Defined Networking (SDN) in the context of AI networking?**

   a. It allows networks to self-optimize without manual intervention
   b. It focuses primarily on generative AI applications
   c. It replaces the need for machine learning models
   d. It is solely used for enhancing cybersecurity

4. **What has led to the rapid rise in the popularity and investment of generative AI (GenAI) models?**

   a. The decrease in computational power requirements
   b. The ability to generate novel text from prompts
   c. The failure of traditional AI systems to meet business needs
   d. The decline in language model applications

5. **Which of the following is a key feature of blockchain technology that ensures the security and immutability of transactions?**

    a. Data can be edited once entered into the blockchain
    b. A centralized authority manages all transactions
    c. Transactions are linked to previous ones, making them tamper-proof
    d. Transactions can be erased if entered incorrectly

6. **What is the core advantage of decentralization in blockchain for secure networking?**

    a. Centralized trust models prevent unauthorized access
    b. Decentralization eliminates the need for middlemen in transactions
    c. Decentralization increases the transaction processing speed
    d. Centralized control ensures data accuracy

7. **Which of the following characteristics of blockchain ensures that data is not manipulated or deleted once entered?**

    a. Transparency
    b. Consensus mechanisms
    c. Immutability
    d. Real-time verification

8. **What is the role of consensus mechanisms in blockchain technology?**

    a. To control and centralize all transactions
    b. To enable nodes to reach an agreement on transaction validity
    c. To eliminate the need for encryption
    d. To increase transaction speed by removing redundant nodes

9. **How does quantum computing enhance AI networking in terms of data processing?**

   a. It eliminates the need for encryption
   b. It processes massive datasets at unprecedented speeds, facilitating real-time data analysis
   c. It slows down AI decision-making to ensure accuracy
   d. It replaces the need for quantum-resistant encryption methods

10. **Which of the following is a challenge when integrating AI with 6G networks?**

   a. Decreased network traffic
   b. Massive data handling and computational power required for AI algorithms
   c. Reduced need for quantum computing
   d. Lack of real-time decision-making capabilities

Answer

1	2	3	4	5	6	7	8	9	10
a	c	a	b	c	b	c	b	b	b

Chapter (05)

THE ROAD AHEAD

5.1 Trends Shaping the Future of AI and Networking

The networking technology is also said to be experiencing a radical change given the following new trends which are changing the whole perception of the connectivity as well processing of data and management of the networks. This evolution is based on artificial intelligence, blockchain, edges computing, quantum computing, and the Advanced Mobile communication while prominently observing the ethical point of view. AI-aided networking is the next paradigm shift in the way networks are configured and guided uses of ML and DL algorithms in the analysis of terabytes of data for configuration, routing protocol tuning, and resource allocation theories. Specifically, blockchain is explaining its importance as the compound for safe networking, as the CIA triad model is used for introducing new safe methods of networking with the help of encryption and safe records. Whereas edge computing is providing solutions to the problem of analyzing huge volumes of sensor data by shifting compute to the edge, quantum computing is providing radical new ways of handling some of the tough computational problems that exist. The progressive enhancement of the mobile communication generation from 1G to 5G and beyond confirms that connectivity standards are jointly evolving and that 5G claims to offer even higher mobile broadband, Light reliable low latency, and massive MTC. Ethical factors play an essential role in these technologies as these systems and solutions evolve, how and where the data is collected and used, conflict of interest, social responsibility, and end user protection. These revolutionary changes are not emerging in a vacuum but are gradually products of integration, and they slowly lay the foundation to a new kind of network that is intelligent, secure, efficient and ethical.

Emergence of AI-Driven Networking

AI-driven optimisation shows promise as a way to improve network efficiency and performance. AI-driven optimisation techniques can analyze large amounts of network data, find patterns, and make wise decisions to optimize resource allocation strategies, routing protocols, and network configurations dynamically by utilising AI algorithms like ML and DL. For network infrastructures to maximize throughput, minimize latency, and guarantee uninterrupted connection, network performance and efficiency must be optimised. In data centers, telecommunications systems, and corporate networks, efficient network management has become a critical factor in delivering the best services while maintaining market leadership in today's digital environment. However, the dynamic nature of contemporary networks and the constantly rising amount of data traffic make it more difficult to achieve optimum network performance.

AI has become a potent instrument for maximizing the effectiveness and performance of networks. The foundations of AI-driven optimisation are covered in this part, along with its definition, ML and DL methods, and applications in network administration and operation. A definition for term AI is therefore the capacity of computer systems to emulate human intelligence. AI refers to a set of methods and algorithms that computers may use to improve network performance by learning from data, seeing patterns, and making smart judgements.

Expert optimisation on the other hand employs integrated computer procedures to sieve through massive amounts of network data for meaningful findings that may enhance performance and efficiency of the network. Networks' efficiency can be enhanced by AI algorithms to perform optimisation work with problem identification, allocation, and route optimisation activities. Machine learning is the subfield of artificial intelligence whose purpose is for a computer to develop its own program via data in order to execute a particular task. These algorithms are able to pick out trends and make conclusions using the data that was used while training it.

One of the most utilized techniques of developing ML systems is supervised learning where algorithms instruct their programs on how to classify new data samples based on previously labelled data. Algorithms can also search for rules or sophisticated shape in a set of unlabeled data by means of unsupervised learning. Bell has categorized ML into the next level, deep learning that uses multi-layered ANN to capture more profound data features. DL algorithms are most efficient dealing with large amounts of data in their original formats, such as pictures and text or even audio.

Numerous problems in contemporary network settings may be solved with the help of AI-driven optimisation, which finds several uses in network administration and operation. Important uses comprise: The use of AI algorithms can track trends in the network traffic, and then adjust the routing protocol in order to reduce traffic and overall latency. Through optimisation methods based on AI, the network resources for example bandwidth might be scheduled to derive optimum performance and effectiveness out of the network resources. It is possible as well that AI systems conduct maintenance activities prior to their breakdown or decline in performance, thereby reducing disruption of the functioning network. Since AI can identify possible security threats and risks such as cyber-attacks or attempts to gain access by unauthorized people, network security may improve with optimisation achieved through AI.

To simplify operations, boost productivity, and ensure dependable performance at scale, AI-Native Networking incorporates AI into the design and development of computer networking systems. The features and advantages of AI-Native Networking are shown in Figure 5.1.

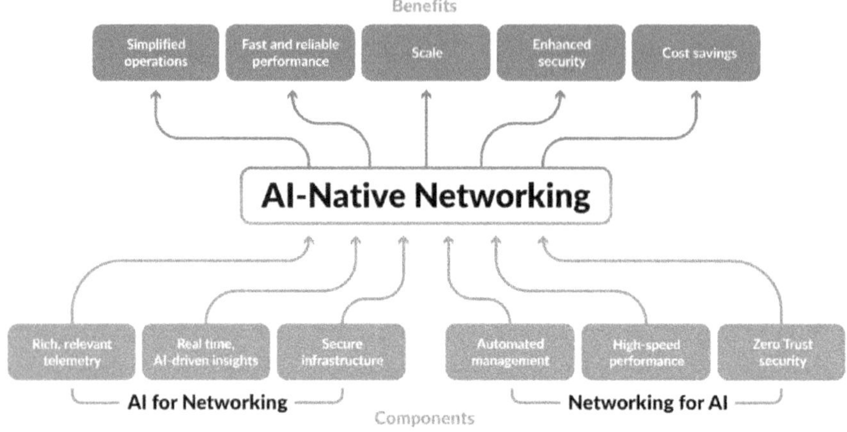

Figure 5.1: AI native networking benefits and components

Source: - (Juniper, 2024)

Application in AI network optimization:

There is a vast variety of uses for optimisation methods powered by AI to boost network efficiency and performance. We will look at several important uses of AI-driven algorithms in this part, such as optimizing routing protocols, network setups, resource allocation, and security measures. Refining network topologies and settings is a key area for AI-driven optimisation in the networking industry. Conventional methods of network design often use static topology and manual setup, neither of which are necessarily suitable for adapting to changing network needs and traffic patterns.

According to certain criteria such as traffic patterns, service requirements or cost constraints, for example, automated network topologies can be generated and optimised through integrated AI-based optimisation algorithms. In order to find the best network configurations for latency reduction, throughput optimisation, and resource efficiency, AI algorithms analyze past network data and forecast future traffic needs. The ability to dynamically reconfigure network topologies in reaction to changing traffic patterns or network circumstances is another benefit of AI-driven optimisation. AI systems can adapt the structure and layout of a network to help accommodate new requirements, as well as managing traffic by always checking for key performance markers and fine-tuning accordingly.

Effective network must therefore call for dynamic routing and traffic management because the traffic as well as demand for network access is not always constant especially in large networks. While network circumstances are always evolving, traditional routing protocols like OSPF and BGP may not be the best choice. The near real-time traffic status, congestion, and performance metrics of a network are also continually amenable to dynamic routing changes since the use of AI algorithms for routing is more adaptive and efficient. AI systems are steeped in the ability to gauge the past and future load patterns on the networks so that it can work out which is the best way of routing the network, way that minimizes delay and maximizes through put and ensuring that the loads are well distributed (Uchenna Joseph Umoga et al., 2024).

This escalating number of threats and vulnerabilities presents significant risks to availability, confidentiality, and integrity of data in a network; rendering network security a critical concern in current networks. Optimisation techniques that employ AI incorporate the use of big data techniques and modern as well as highly advanced machine learning techniques to find new ways of enhancing the efficacy of network security and threats identification. Network traffic in real-time might be managed by AI-based algorithms, detect malicious activity, and may identify security threats such as virus, intrusions, and DoS attacks (Negnevitsky, 2005).

Blockchain application for Secure Networking:

The CIA triad model is the standard for developing an organization security model for networks in the field of computer security and network security. The triad consists of three components; availability, integrity and secrecy. We can verify compliance with all of these rules using blockchain technology.

- **Confidentiality:** The last strategy of protecting sensitive information is to minimize the access to it through only those who retain legitimate right and need to view it. Data entered on a blockchain cannot be observed by illicit persons even when transferred via unsecured networks due to high encryption. To reduce attacks on the network, it important to put adequate security measures like for application-level access. Applying critical public information systems, blockchain technology embeds identities and secures the dialogue, which may improve organizational security. While secondary storage is convenient for storing backup copies of private keys, it greatly increases the likelihood that these keys may be lost or stolen. To prevent this, it is recommended to dwell on cryptographic techniques tied to integer factorisation problems and key sophisticated management systems such as Rfc or IETF.

- **Integrity:** Blockchains help organisations secure the integrity of their data because to its inherent qualities, such as immutability and traceability. If a cyber control assault were to use 51% of the network's resources, consensus model protocols might help organisations even more with ledger splitting security and management. Here, at the beginning of each new cycle, the current status of the system is saved in the Blockchain, creating a fully immutable history log. To control data mining blockages, smart contracts may be employed to check and even enforce the agency's standards with such participants.

- **Availability:** DDoS assaults have become more common as of late, and their goal is to interrupt the availability of technical services. The hacker attempts to overload the networks with a vast volume of relatively tiny transactions, making DDoS assaults costly in blockchain-based systems. There is much less chance of IP-based DDoS assaults impacting the overall operations of a blockchain since it does not have a single point of failure. The data becomes available in several nodes to guarantee that there is a full copy of the ledger pulled through. The distributed integration of several nodes and processes makes the systems and platforms robust (Rajendran et al., 2023).

Quantum Computing and AI

Quantum computing technology makes it possible to solve problems in a way that is both more effective and fundamentally different from what is possible with traditional calculations. Additionally, it provides whole different solutions to computational problems. Positive experimental results point to the possibility of commercially available quantum computers within the next several years. Shor's prime factorisation technique is regarded as one of the most significant and well-known examples of an algorithm based on the capacity of quantum computers. The quantum computer may possibly solve the computational task in a few hours, while a conventional computing environment would take billions of years. The "great boom" in the area of quantum computations in 1994 was caused by this method.

Classical information theory's findings are insufficient to characterize communication across a quantum channel; instead, quantum perceptual generalisation is necessary. An encoder and a receiver participate in the standard model of communication over a quantum channel Q, with the former encoding the message and the latter decoding it. Here, however, a quantum system is responsible for the whole communicating process. The encoding process in quantum channels is radically different from that in conventional encoder schemes because the information travels via quantum states. Encoding in this context refers to the process of preparing a quantum system in accordance with the probability distribution of the classical message that is being encoded. There is also a change in the decoding procedure, which here refers to the measurement M of the received quantum state. In order to understand the Q channel's characteristics and the key distinctions between it and the classical channel, one must have a foundational understanding of quantum information theory. All possible physical changes on the transmitted quantum system are shown in the quantum channel model. Matrix representations of quantum states are produced by the channel transformation, which in turn produces another quantum system. Despite their diversity, all physically permitted channel transformations are CPTP transformations, where "trace" is defined as the sum of the components on a matrix's principal diagonal. Consequently, the trace preservation characteristic ensures that the density matrices (a mathematical representation of a quantum system) at the channel's input and output have identical traces. An information-encoding quantum state S is fed into a quantum channel as input. A quantum communication channel, which may be realized in reality by optical fiber channels or wireless quantum communication channels, is used to transmit the quantum state. The receiver must take a measurement in order to deduce any information from the quantum state. The result of the perturbed quantum state measurement whose quantum channel transformation determines whether the result is completely probabilistic or deterministic. One difference between classical and quantum channels is that the former converts encoded data into quantum states. These states may be anything from the particle's spin to the ground and excited states of an atom, or any number of other physical ways, as discussed in Chapter 8. The classical capacity of a quantum channel becomes important when classical information in a quantum state or private classical capacity via quantum systems needs to be sent. Quantum entanglement and superposed quantum states are examples of quantum information that could be transmitted via quantum capacity (Glisic & Lorenzo, 2022) optimization, and dynamic control. Machine learning algorithms are now often used to predict traffic and network state in order to reserve resources for smooth communication with high reliability and low latency. In Artificial Intelligence and Quantum Computing for Advanced Wireless Networks, the authors deliver a practical and timely review of AI-based learning algorithms, with several case studies in both Python and R. The book discusses the game-theory-based learning algorithms used in decision making, along with various specific applications in wireless networks, like channel, network state, and traffic prediction. Additional chapters include Fundamentals of ML, Artificial Neural Networks (NN.

5G and Beyond

Mobile communication technologies have transformed our lives. They remove time and space constraints for transferring information from one location to another. This enables anyone to share or access information anytime, wherever, and however they choose. Mobile communication systems are become a vital element of our daily life. Over many decades, mobile communication methods have changed. We must look back on our progress from the ground-breaking IG to 4G and consider the significant changes from generation to generation in order to completely comprehend where we are in 5G (Fig. 5.2).

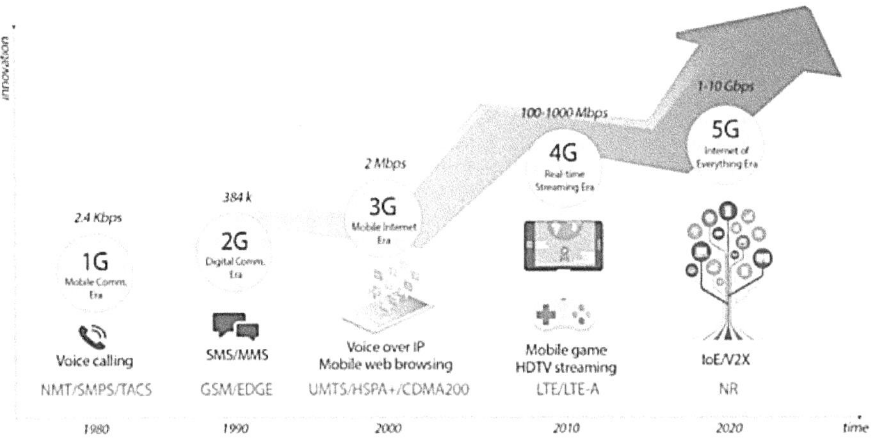

Figure 5.2: Evolution of Mobile Communication Standards

Source: - (Lin & Lee, 2021)

The needs of consumers for mobile phone and broadband data services were mostly met by earlier generations of mobile communications networks (1G to 4G). The next mobile communications technology is called 5G. It expands upon the achievements of earlier iterations of mobile communications systems. 5G promises to enhance the end-user experience and open up new income streams, ecosystems, and services. More effective spectrum utilisation, larger capacity, reduced latency, and much faster data rates are all things that 5G is said to provide. Due to its enhanced capabilities, 5G is capable of supporting a wide range of use cases and applications. Three types of possible 5G network use cases have been identified by ITU-R: eMBB, URLLC, and mMTC.

The ITU Runs a program called IMT-2020 to build international mobile telecommunications for 2020 and beyond. It began in early 2012. The detailed specifications for 5G's technical performance are laid out in IMT-2020 and may be found in Table 1.1. With 5G, data speeds will be ten times higher than 4G. Uplink (UL) peak data rates of 10 Gbps and downlink (DL) peak data rates of 20 Gbps are anticipated, respectively, in 5G. At the five percentiles, SG can offer user-experienced data rates of 100 Mbps in the DL and 50 Mbps in the UL in crowded metropolitan environments. 5G will provide a vast improvement over current mobile broadband due to its much higher data speeds. It is anticipated that 5G spectral efficiency would be much higher, with peak DL spectral efficiency reaching 30 bps/Hz and UL spectral efficiency reaching 15 bps/Hz. Multiple eMBB settings can accommodate varying user spectral efficiency. A good illustration of this is the fact that 5G can provide DL and UL user spectral efficiencies of 0.12 and 0.045 bps/Hz, respectively, in rural regions, and 0.3 and 0.21 bps/Hz, respectively, in indoor hotspots. Thus, 10 Mbps/m^2 is the maximum capacity for DL area traffic in an indoor hotspot. In comparison to 4G, 5G spectral efficiency is three times better. The anticipated 1 ms user plane latency from 5G is a tenfold reduction over 4G's user plane latency. The much-enhanced latency performance is capable of satisfying the demanding latency standards of autonomous transportation and the industrial internet of things. Furthermore, 5G guarantees much better dependability and performance. We anticipate a 99.999 percent success rate in sending a 32-byte layer 2 protocol data unit within 1 millisecond in a macro-urban setting with good channel quality and coverage. There will likely be a considerable improvement in mobility performance with 5G. Time spent travelling will be interrupted by zero milliseconds. 5G enables service quality that is sufficient even at mobility speeds of up to 500 km/h. An anticipated density of one million devices per square kilometer is being accommodated by 5G. When it comes to mMTC situations, the increased capacity of 5G is perfect. Energy use is another important factor to think about. High sleep ratio and lengthy sleep duration support are anticipated to contribute to 5G's network energy efficiency.

The research community have now begun to consider the future of mobile communication systems, often known as 6G systems, since the 5G NR standard is coming into its own. While 6G is still in its infancy, this section will attempt to provide an overview of its goals, obstacles, and essential supporting technologies. This overview offers some unique perspectives on 6G while sharing some similar themes with. 6G: what exactly is it going to be? In theory, consumers should be able to access intelligent, personalized, or task-dependent services from any location using 6G networks. The world will soon be populated with billions of wireless gadgets, sensors, and mobile phones, all over the place. These gadgets will actively seek for network connectivity whenever they feel the need. By communicating with (distributed) data centers, each of which has powerful processing capabilities, this generates meaningful information for each specified activity. Consequently, 6G will revolutionise wireless communication by linking all wireless devices and promptly giving consumers with optimised information based on their individual needs in every given setting. One potential goal of 6G is the introduction of new intelligent connection services (Lin & Lee, 2021).

Ethical AI in Networking

Loosely linked computing parts and other devices form a dispersed system known as a computer network. In this setup, any two of those devices may talk to each other across a communication channel. The channel could be either wireless or hardwired. Distributed systems cannot be termed networks unless they adhere to a predetermined set of rules for communication known as protocols. These guidelines must then be adhered to by every network communication device in order for it to interact with other devices. It looks like a typical wired computer network. Each network node may be the owner of either local or global resources. These resources might have a hardware or software foundation. Network protocols and application programs that are used to synchronize, coordinate, and facilitate data transfer and exchange across network nodes may be considered software. Additionally, network software allows costly resources to be shared inside the network.

Online social communities differ greatly from conventional physical social groups in that they lack an epicenter of authority, to which all members must submit with a feeling of shared duty. Since this kind of community governance is new and involves equitably distributed power and accountability rather than a central command, a system must be established and adhered to in order to protect each and every community member. However, these processes have not been fully described, and even in cases where they have been, it is premature to draw any conclusions about their efficacy. These factors amplify the already high levels of complexity, unpredictability, and decentralization:

- ***Virtual personality***: You are familiar with their names, as well as their preferences. You don't know them at all, yet you know them so well that you can even wager on their thoughts. They are impossible to meet and identify in a crowd.
- ***Anonymity:*** They are a constant presence in your professional life. Despite the fact that you know them just by a first name, they are your pals. Both you and they will always stay in the dark about who they are.
- ***Multiple personality:*** You believe you know them, but you don't since they might change and evolve into other people. They are capable of assuming as many different personas as there are topics under discussion. It is impossible to predict which personality you will encounter next. The frequency of young suicides linked to cyberbullying is increasing, as these figures show.
- ***Risk for child safety:*** Cyberbullying and site abuse are not the only issues with online social networks; there are also actual risks to children, whether or not they are being cyberbullied. Child exploitation in online social networks is on the rise. According to the most recent statistics, 600,000 kids use Myspace, while around

1 million youngsters under the age of 16 use Bebo. The risk of internet child abuse is increasing with these figures. Although allegations of abuse often reach the site administrators instead of the authorities, the networking sites claim to be enabling users to do so. Efforts are being made by governments worldwide to have a better understanding of the issue and identify potential remedies.

- ***Psychological effects of online social networking:*** There has been a meteoric surge in both the quantity and quality of online social networks due to the proliferation of both users and membership in these platforms. As the number of users increases, so does the number of people experience issues. Experts have begun to recognize Internet addiction as a distinct mental health condition due to the increasing number of people—particularly adolescents—who spend an unhealthy amount of time online, particularly on social networking sites. The most prevalent indicators of social media addiction, as stated by Neville Misquittaa in "Psychiatry and Society in Pune," are: People who are naturally outgoing and confident are more likely to get addicted to social media. Facebook is popular among shy individuals, and they tend to spend more time there. They do not, however, have many Facebook "friends." High amounts of online social involvement are also associated with narcissistic personalities. They are identified on the internet by the volume of their social media engagements, the beauty of their primary picture, and the self-promotion of their primary photo.

- ***Free speech:*** When using an online social network, what kinds of expression are protected? Despite the fact that the National Labour Relations Act protects employees from being fired for "protected concerted activity," which forbids employees from being fired for collective action but permits employers to fire employees for individual actions they take against the company, the issues surrounding online social networking remain unclear, and it is still unclear what kinds of speech are protected in these platforms (Kizza, 2016).

5.2 Global Impacts of Intelligent Networks

5.2.1 Transforming Industries with AI

- **Healthcare**

Testing a person to detect any changes in their health status due to exposure to certain environmental health dangers, such as airborne pollutants such harmful dusts, vapors, or fumes, is known as health monitoring. In recent years, wearable computing and sensor technologies have emerged as a cutting-edge and popular way to enhance patients' quality of life in the healthcare industry. An important part of healthcare monitoring systems like IPCS is the integration of wearable and ubiquitous sensors with personal mobile devices. This allows for the continuous monitoring of patients as they go about their regular lives, which may help avert catastrophic circumstances. Health care providers often use wireless technology and mobile devices for bio-monitoring and home monitoring. The field of bio-monitoring makes use of mobile networks to track vital signs such as blood pressure, temperature, oxygen saturation, ECG, and heart rate. Various sensor and mobile technologies may be used to gather data that takes the patient's environment, including indoor and outdoor temperatures, location, and activity status, into account. The network is networked, allowing medical personnel to access patient data, test results, and medication information from anywhere. As shown in Figure 5.3, IPCS offers a broad variety of services to the general public and medical professionals, including location-based services, telemedicine, incident detection, emergency response and management, patient monitoring, and ubiquitous data access.

- The goal of patient monitoring is to employ sensor technology to remotely observe patients' health states. Healthcare practitioners may then access the data to see the patient's current state of health.
- The ability to read and update clinical information using portable devices is made possible by ubiquitous access to data.
- Using mobile devices, telemedicine allows medical professionals, including physicians and specialists, to diagnose and propose treatment in both routine and urgent conditions from any place.

- Healthcare providers are promptly notified via incident detection in the event that a patient's health state changes.

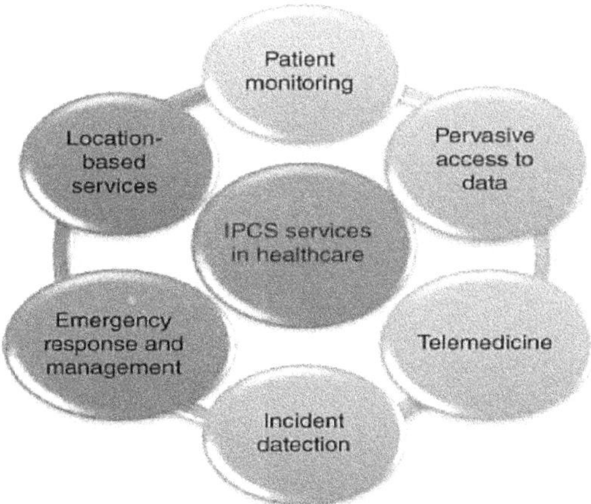

Figure 5.3: The Role of IPCS in Modern Healthcare
Source: - (Sangaiah et al., 2019)

- The public is effectively served by emergency response and management, which sends nearby ambulatory vehicles to the scene of the incident and directs patients to hospitals in the area.

- Healthcare personnel are able to better oversee elderly or mentally impaired patients who need mobility care but are limited to certain places thanks to location-based services. In case of an emergency, this tool also aids users in locating nearest medical services.

- **Healthcare System Architecture Based on Monitoring**

An environmental monitoring-based healthcare system architecture is described in this section. Wireless sensors (or biosensors), monitoring equipment, gateways, e-platforms, computer devices, and cloud infrastructure are all part of it. For effective handling of data from diverse devices, the system employs the idea of cloud computing. Cloud computing enables ubiquitous access to data and resources as well as direct sensor-to-infrastructure connectivity and shared infrastructure. The six levels of the layered architecture are as follows: physical, mobile, interoperability, emotional, context-aware, and application. Each layer is detailed in detail below and is shown in Figure 5.4.

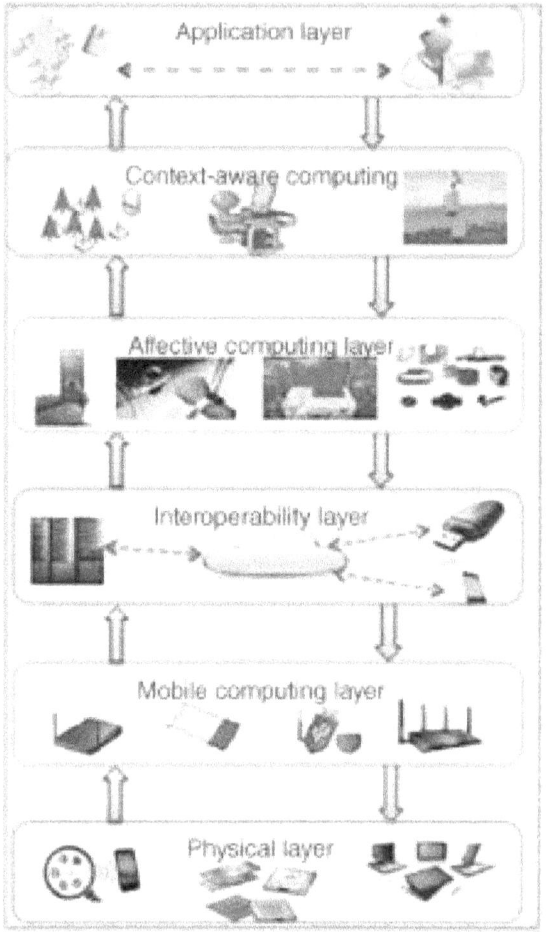

Figure 5.4: Healthcare system architecture based on monitoring

- **Physical layer:** The initial level of this design is the physical layer, which houses the many pieces of hardware needed to carry out intelligent ubiquitous computing. In order to do computations and monitoring, the heterogeneous hardware comprises a CPU, input/output devices, sensors, and a network interface card (wired or wireless). This monitoring data will be helpful for making judgements about environmental health and preventative measures, and all the heterogeneous devices are linked together by wired or wireless technology. Either a real person or a program may transmit and receive signals in a ubiquitous computing system. Every entity in the physical layer gets along with every other entity.

- **Mobile computing layer:** Layers physical and interoperability are sandwiched between mobile computing. There is a greater degree of abstraction here than in the physical layer, and it specifies the device resources. It employs communication, computation, and portable devices to gather data from various distant places, which can be accessible by any user, anywhere. In centralised mode, sensors provide data to a server, which then summarises and processes the information. Machine learning, decision-making, spatiotemporal reasoning, fuzzy rules, neural networks, and other data mining and AI approaches are used to analyses sensor data from various environments.

- **Interoperability Layer:** When it comes to mobile and affective computing, the interoperability layer is middle ground. In this layer, data from various sources is made available to users or service providers via the cloud server in a resilient and scalable manner. It is simpler to get data, manage it, and communicate with other apps when you use cloud infrastructure. Database services, storage for queues and tables, web roles in the cloud, and traffic manager are some of the cloud computing services used for data management. The consolidation of data and the enhancement of available resources may enhance system security.

- **Affective computing layer:** The affective computing layer sits between the context-aware computing and interoperability layers. Resolving and detecting the impacts of multimodal input is its primary objective. It stores a variety of environmental sensors, including those that measure temperature, humidity, wind speed and direction, greenhouse gases (GHGs), and more. The environmental pollutants' values, collected from various sensors, are stored in this layer, and they are sent to the context-aware computing layer for processing and subsequent decisions.

Intelligent systems, which include biosensors, wearable sensors, and portable devices for communication and processing, enable human-computer interaction. In preparation for further processing, data is integrated and analyzed from a variety of sensors.

- **Context-aware computing layer**: Decision-making is primarily facilitated by the second uppermost layer of this design, which is the context-aware computing layer. The affective computing layer transmits information about pollutants, which this layer uses. Information on the user's location, activities, and other contextual factors are provided at a high level. Health care practitioners and consumers will have access to automatically created, situationally relevant information about environmental contaminants. Service delivery to the authenticating user and network resources occurs at this tier. In addition to providing data needed to update the network, it is in charge of connecting and disconnecting resources. Government organisations may use this layer to assist them establish pollution-related rules and deliver the right services.

- **Application layer:** The architecture culminates with the application layer. It is on this layer that a number of useful apps are stored. The healthcare system is built within the ubiquitous computing architecture's application layer. A user-handheld device-specific interface component for ubiquitous application is located in this layer. The component gives the user or developer the ability to choose the desired communication method and construct the necessary user interface. It incorporates the use of communication technologies to provide healthcare providers with real-time information on patients' health condition via continuous patient monitoring. Changes in ambient pollutant levels will trigger alarms to be sent to the application layer based on information from the context-aware computing layer. By informing patients via mobile devices or wireless sensors, healthcare providers may take necessary precautions and procedures to mitigate hazards.

In order to create a scalable, flexible, and heterogeneous environment, the architecture can make use of various wireless networking and sensing technologies, such as on-body sensors with long-range radios, short-range BANs, on-body sensors with a long-range gateway, inside-body sensors with wireless communication and power delivery, inside-assistive device sensors, outside-sensors, and, of course, those ubiquitous mobile smartphones.

• Manufacturing

The inventory management system's many IIoT-based applications have led to the digitization of business and industrial processes. For instance, RFID tags find the objects, scan them, and send the server the item's specific details as an inventory transfers from one unit to another. This real-time inventory information may be seen remotely by the inventory manager.

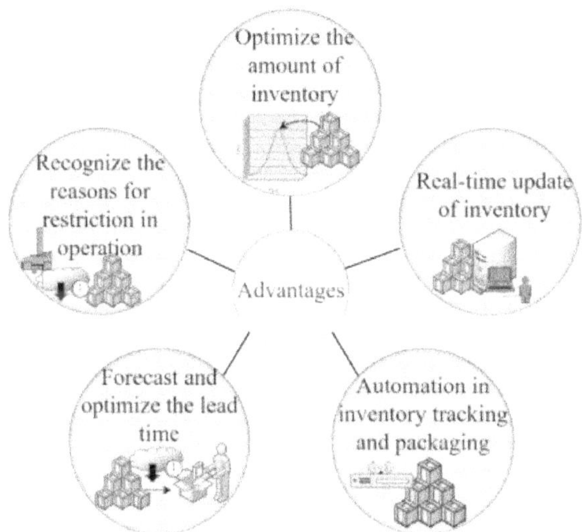

Figure 5.5: Network-Based Inventory Management in Continuous Manufacturing
Source: - (Goundar et al., 2021)

The following are some advantages of incorporating IIoT into inventory management:

- **Automation in the process of inventory tracking and packing:** The inventory may be monitored in real-time with the integration of RFIDs and other smart sensors based on the IoT (Fig. 3). Plus, these records are kept in databases on distant computers. As a result, manual processes are made easier, human mistake is less likely, and up-to-the-minute stock information is preserved.

- **Real-time update of the quantity of inventory items, their location, and tracking:** The inventory management is constantly updated with real-time data on completed goods, remaining raw materials, and work in progress thanks to the upgraded tracking system. The inventory manager is also kept informed of the whereabouts, status, and movement of products inside a warehouse.

- **Optimization of the amount of inventory:** The inventory manager can better control the placement, quantity, and timing of inventory goods with real-time data on stock levels.

- **Recognize the reasons behind bottlenecks in operation:** The manufacturer is able to identify machines with lower operating efficiency and the quantity of buffer and anticipatory inventory that has to be kept since the inventory management gets real-time information about the location and state of the inventory.

- **Forecast and optimize the lead time:** The information on the quantity of stock on hand and the ability to predict the warehouse's buffer stock are both held by the inventory management (Goundar et al., 2021).

- **Transportation**

Vehicular networks provide a variety of services and applications to its customers when it comes to vehicle communication with other cars and the road infrastructure. While traffic safety was the original motivation for vehicle ad hoc networks (VANETS), additional issues surfaced. The many application kinds fall under the following categories:

- **Safety** refers to the category of apps designed to improve driving safety by transmitting and receiving relevant information about public roadways. As a result, these application classes have the ability to spread knowledge about an incident that happened on a road or street. Drivers may find this information instructive, or it may trigger roadside signs. Cooperative crash alerts, emergency video streaming, emergency warning systems, and others are a few examples.

- **Comfort** is the category of applications, like traffic information systems, that seek to optimize routes, improve passenger comfort, and increase traffic efficiency. exchanges amongst car occupants via voice messages, chats, music downloads, and other means.

The primary objective of safety applications is to decrease the frequency of vehicular accidents. Because of the time-sensitive nature of these applications, notifications must reach vehicles or drivers in sufficient time to enable them to take precautions to prevent accidents. Data and vehicle-to-vehicle communication strategies are used to decrease message delays. Here are some of the most researched and developed applications in the automotive industry and academia:

- **Post-Crash Notification.** An automobile engaged in an accident is the setting for this kind of application. In order to notify other cars of its whereabouts, the impacted vehicle starts utilising data dissemination systems to spread alert messages to them. This message propagation is done in order to promptly notify other cars about the collision, allowing them to make choices and request a rescue team. For applications in this class, both V2V and V2I communication can be considered as feasible solutions. In order to enable first responders and other cars in the vicinity to take the necessary action, vehicles may thus spread information about an occurrence. Reducing the incidences of false positive and false negatives is another advantage of such communication which stops wrong information concerning an accident from circulating.
- **Cooperative Collision Warning.** Drivers are alerted about the possibility of an impending accident with another vehicle by this sort of notification. As a result, other cars will be able to respond to prevent a collision. In order to prevent both drivers from doing the identical actions to avoid a collision, this kind of application

requires a collection of sensors that can detect when another vehicle is approaching and analyze the driver's behaviour.

- **Lane Change Assistance.** A driver's actions while merging lanes or passing other cars may be tracked by this kind of software. To let other cars in the area, know about a lane change, a message is sent out across the network. Both active and passive modes may be used with these apps. The vehicles that are engaged in monitoring changes in movement calculate their distance from each other in passive mode. Active mode is a process of sending signals to other cars in order to find out how near they are and whether they are changing their behaviour or not. The message spreads to other cars in the vicinity when a lane change and direction change occur.

- **Road Hazard Control Notification.** This means cars communicating with each other about an obstacle like a landslide or approaching change in car orientation like a sharp turn or hill descent.

- **Traffic Vigilance.** This method is in the form of a surveillance technique to monitor behaviour of motorists on roads and highways and may well help prevent road rage and other violations through a system of punitive measures. The aim is to verify and identify dangerous driving behaviour.

5.2.2 The Role of AI Networking in Economic Growth

AI is emerging as the financial markets' complex algorithm and machine learning methods that enable the evaluation of massive data and making the right decisions. AI in the context of the financial markets is the use of sophisticated computer methods to the automation of tasks, identification of patterns, and optimisation of results. Due to the potential for AI to help in enhance financial system in developing countries where the infrastructures have not well developed this is especially important. This introduction will try to explain how AI could help to solve problems that are pertinent to emergent economies as they struggle to enter financial markets. While authors continue to acknowledge the importance of AI for financial market evolution, the stance of AI in emerging economies lacks literature. Ironically, although many investigations have been made to examine the trend and impact of AI in advanced economies, little systematic analyses have paid attention to investigate the strategies and benefits of AI applications to cope with the emerging economies' issues.

Emerging economic countries present a spectrum of financial systems from highly developed ones to the ones that are still in their infant stages. They play an important role in most of these economies as they act as a link between savers on the one hand, and borrowers and investors on the other hand. Another tough managerial test of developing economies is the problem of weak access of credit facilities which is affecting most SMEs and most especially the rural population. Unfortunately, this kind of credit access is very limited, which in turn slows down entrepreneurship, investment, and overall economic improvement. Other challenges include frail institutional environment that discourages investors, poor legal structures that fail to support sound markets, and ineffective institutions that deter growth in transparent markets. Technology also adds to these difficulties because of its constraints. There is a problem of outdated infrastructure, low levels of internet connection, and low levels of digital literacy which offsets the adoption of advanced digital financial services in many emerging economies. Consequently, many people still don't have access to official financial services and must depend on less secure and more informal methods of financial intermediation. Further, growth in emerging market Financial Systems is sensitive to external shocks, and financial crisis by factors such as; fluctuation in capital, currency, and political instability. These vulnerabilities sap investor confidence, trigger capital flight and deepen downturns further encumbering the development of financial markets. Solving these issues entail promotion of innovation and technological progress in the sphere of the EMS's financial systems. Through the latest technologies ranging from artificial intelligence, block chain and digital payments, new economies can work towards achieving higher productivity, integrity and financial inclusion. For example, millions of citizens who cannot obtain any banking products at the moment can get them cheap and fast through such systems as digital payments and mobile banking, which means a full-fledged revolution in the sphere.

A good network base and innovation culture are some of the key components of a well-functioning and wealthy economy. Over fifty percent of the world's population is living in cities making it much more important to enhance mass transportation systems and develop new types of renewable energy sources. The authors Conde and Twinn continued by saying that new sectors, advances in ICT, and artificial intelligence are crucial due to the enormous increase in urban populations. Another incontestable truth is that, in the poor world, more than 90% of the world's 4 billion internet users do not have access to the internet. A combination of increased investment in infrastructure, wider availability of information and knowledge, and encouragement of AI-driven innovation and entrepreneurship is necessary to close the digital gap. Furthermore, infrastructural restrictions reduce company productivity in many low-income African nations by around 40%, and it is thought that 2.3 billion people worldwide lack access to basic sanitation. Also, according to the UN, some 2.6 billion people in developing countries do not have access to reliable power, and the renewable energy industry is now employing over 2.3 million people, with that number expected to increase to 20 million by 2030. The fact that only about one-third of agricultural goods in developing nations underwent industrial processing, in contrast to almost all of those in high-income nations, is another major issue that arises from inadequate infrastructure.

Global economic production might increase by almost $13 trillion by 2030 as a result of AI advancements, say Conde and Twinn. AI is already having an impact on the global economy. From 2017 to 2023, the worldwide market for AI technologies connected to transportation was $1.2 to $1.4 billion, and experts predict it will reach $3.1 to $3.5 billion by 2023, representing a rise of 12 to 14.5 percent. Several sectors may benefit from AI interventions that enhance infrastructure; for example, the transport and agricultural sectors can be transformed to work towards the SDGs. According to the World Bank, AI has several potential uses in transportation, including the development of autonomous cars and the resolution of other issues pertaining to pedestrian and vehicular safety. Particularly in less developed nations, road traffic accidents pose a threat to public health. According to Conde and Twin, the number of fatalities caused by traffic increased to 1.35 million in 2016 from 1.25 million in 2013, with the majority of these incidents occurring in countries with low economic levels. Inadequate infrastructure, bad roads, and cars without current safety equipment are among the factors mentioned, along with human mistake.

In the European Union (EU), almost 90% of all traffic accidents are caused by human error, including speeding, distracted driving, and intoxicated driving. In the EU, these incidents claimed the lives of almost 25,000 individuals in 2017 alone. According to research, in some industrialized countries, the use of AI in driverless cars might help cut traffic accidents by over 90% by 2050. The instant self-driving technology were engaged, Tesla's first effort at an autonomous car decreased accident rates by 40%. Although there may be some advantages to autonomous vehicles, it will take a while for them to be widely used in developing nations. However, according to some academics, by 2030, there will be one driverless car for every four autos.

- **Key ways AI networking contributes to economic growth:**

  - **Faster Innovation:** Businesses may speed up their innovation cycles by creating customer-centric new goods and services with the help of AI-powered network analysis that identifies trends and patterns in data consumption.
  - **Increased Operational Efficiency:** With dynamic changes in the settings of a network and analyzing the network information in this regard with the help of AI algorithms, there can be prediction of probable issues and handling of bottlenecks, thus reducing the downtime and increasing total efficiency of the business in the network.

- **Enabling New Business Models:** AI is particularly good at reimagining a business value chain, based on the real-time analysis of data and automation, which means that new opportunities open up in IoT, cloud computing, and edge computing markets.

- **Cost Reduction:** Network maintenance and troubleshooting operating expenses may be drastically reduced with the use of AI by automating mundane operations in network administration and optimising the allocation of resources.

- **Examples of AI applications in networking:**

  - **Network Traffic Management:** Algorithms powered by artificial intelligence can optimise network bandwidth allocation for mission-critical applications by responding to real-time traffic patterns.

  - **Security Threat Detection:** AI has the capability to analyse network data in real-time, allowing for proactive security measures by identifying possible threats and abnormalities.

  - **Self-Healing Networks:** AI can automatically detect and resolve network faults, minimizing downtime and improving network resilience.

  - **Predictive Maintenance:** AI can analyze network device data to predict potential hardware failures, allowing for preventative maintenance to avoid disruptions.

5.3 Challenges in Scaling and Implementation

Infrastructure Limitations

This is true when it is a question of scaling any technological solution that is implemented in a fix asset environment because when it is a question of adding new viewers, the challenge is normally very much infrastructure-related. The first challenge is the lack of equal access to connectivity across different locations of the world. For instance, unstable internet connection may be a result of geographical location and or can be the result of a low level of development hence restricting. The implementation of complex systems. In addition, there is a realization that computational demand grows quickly as the system sizes increase, and this requires efficient computing hardware and software. This is an even more significant problem in computing-hungry applications like machine learning and artificial intelligence where there is high demanding for resource use.

Closing these gaps entails the allocation of resources in infrastructure development bearing key strategic factors. Both government and private sectors have to come together especially in an effort to extend broadband infrastructure and to establish appropriate data centers. Information cloud can actually afford solutions for cop processing requirements depending on the demand. However, such approaches require a lot of capital at the beginning of the method implementation and constant support in the future.

Another factor that has to be thought through is energy use. However, as computational systems scale up, such energy demand supplements existing power structures and tends to have high costs. Concerns include high costs of ICT, rapid obsolescence and scalability, and data security and privacy issues of cloud computing can be addressed by promoting green computing technologies and energy efficient technologies and hardware. In the end, improvement of limited infrastructures is a combination of new technologies, funding commitment as well as policies.

Data Privacy and Security

Currently, nobody can overemphasize the importance of data security and privacy in the age of the Internet. Through scaling technological solutions, it is important to manage large quantities of data, very often consisting of sensitive information, therefore the question of protection, as a result, becomes much more critical. Other threats, like leaked data, unauthorized access and cyberattacks, are extreme threats which lead to trust loss and have outstanding legal and financial implications.

Security of data is an essential step and if a good start is made, through practices such as using encryption protocols, then the next step is taken.

Further, anonymization techniques can guarantee that definite information will never be identifiable. They range from performing security audits at least once a year; penetration testing; and ensuring compliance with international best practices act, GDPR and HIPAA acts.

Continuity of informing innovative ideas and eradicating threats to customers' privacy is another issue. Public and private organizations need to be open about the way they gather data, how they store it, and how they use it. Trust is built beyond regulatory measures; it needs to empower the user to own the data management process. New technologies like blockchain, can also have great value as new ideas and solutions for handling data in a secure and decentralized manner.

Interoperability Issues

Technological ecosystems are growing, and thus getting various systems have to remain unified becomes very important. The causes include lack of standard protocol, dissimilarities in the software structure, and proprietorial software systems that do not support integration.

Lack of LO Interoperability results to poor performance standards, redundancy of processes, and all these makes the user experience disjointed. For instance, in the healthcare industry, electronic health record (EHR) systems by different organizations vendors do not integrate, hence complicating issues such as patient flow and data exchange.

Solving these problems needs the enhancement and implementation of global norms. Developing such standards must be initiated by industry wide collaboration in order to define and create solutions compatible with each other and encourage innovation. Web services and APIs are highly useful for making technologies compatible and turnkey methods that can integrate with other systems. Moreover, the professionals working in training need to be put into knowledge about the interoperability solutions. To counter these barriers organizations should consider being able to develop skills and also promote collaboration. A greater focus on the aspect of interoperability in systems implies that they can be made more efficient, scalable and satisfying to users.

Regulatory and Ethical Hurdles:

Managing the legal and ethical concerns is one of the challenges encountered when growing technological product solutions. Regulations on technology differ from one country to another and this is something that hampers work in organizations that are present in different countries. Staying within these frameworks while still being creative, however, can be a real challenge. The two concepts merge most of the times depending on the kind of business or company under consideration. For example, such important ethical objectives are technology neutrality, Non algorithm bias and protection of vulnerable clients or communities. Concerning these issues public trust is highly affected depending on how the concerns are handled. Horrifically, any organization that cannot manoeuvre these obstacles stands to face public backlash, legal suits and even spoil their reputation. For that reason, organisations should independently seek channels through which to communicate with policymakers, industry members, and the public at large to influence the design of policies that are inclusive of the characteristics of organisations as well as acknowledge organisational transformations. Ethical principles should be introduced into the process of creating technologies in order to improve matching between them and effective values. Likewise, it is possible to assess the impacts of processes and activities, as well as to involve other organizations in auditing the risks involved. Education, as well as awareness creation is also crucial. When the public is better informed and educated about the application of the technology, the organizations implementing the technology find it easy as the public is ready to accept the change. Finally, dealing with the regulatory and ethical problems is not just the compliance question; it is the strategic question that helps achieve sustainable growth.

Cost Barriers:

High initial investments in infrastructure, research, and development limit accessibility for smaller organizations and developing region.

Entrance costs involved in development of networking infrastructures, and especially research and development costs, act as major barriers towards the implementation of AI networking solutions. The development of these structures like data centers, edge devices, and high-speed networks is capital intensive and for small institutions and enterprises in the developing world is costly. In addition to that, it's expensive to acquire the state-of-the-art methods such as GPUs, TPUs and AI Accelerators.

Apart from the costs of deploying AI capable equipment, the costs of developing and customizing AI resident networks require considerable investment – the license fees for AI frameworks and integration costs of AI into contemporary network infrastructures. Substantial growth in AI means that extended research and development is required, which also added to various operational costs such as hiring the best talent to work with AI and cybersecurity professionals, networking experts.

Additional challenges faced only by organizations in developing regions include undeveloped telecommunications networks, unreliable power supply, and high costs of importing efficient technology. Additionally, to meet the international data protection laws and cybersecurity also cost a lot of implementation costs especially for the multinational company willing to operate more than one country.

Whereas, large scale enterprise might be able to bear such costs as strategic expenses, the FOSS community and the small firms and start-ups suffer from the challenge of raising sufficient capital to deploy these disruptive technologies. Allowing people to build a connection towards having AI empowered networks entails using public private partnerships to employ the government incentives, provide a platform through which open-source frameworks on Artificial Intelligence could be developed.

5.4 A Vision for the Next Era of AI Nexus Systems

The next generation of AI Nexus Systems aspires to the proactive synergy of such technologies as Artificial Intelligence, IoT, and human-oriented design to form smart systems and networks. Closely related to this vision, some of the key ideas include: ubiquitous intelligence which places use of embedded systems for real-time intelligent decisions; human centric AI, which deals with improving humans while incorporating ethical values like justice, dignity and creativity; resilient and adaptive networks which aims at creating intelligent networks that are tolerant to faults, robust and efficient in dynamic environments.

Additionally, collaborative AI ecosystem have embracing human-AI partnership, fostering roles such as AI trainers, ethicists, and interaction designers that embrace the link between artificial intelligence and society, business, and government requirements. These systems pertain to the positive integration of automation with the creativity and wisdom of the human mind to revolutionalize particular industries as well as the global society as a whole while achieving the highest standards of ethics and flexibility according to the ever emerging and developing trends in technology.

Towards Ubiquitous Intelligence

The term UbiCom, which combines the words "ubiquitous" (meaning "appearing or existing everywhere") and "computing," describes ICT systems that make data and tasks accessible everywhere, facilitate intuitive human use, and are invisible to the user. The term "ubiquitous computing" (UbiCom) refers to the widespread use of computers in many contexts, not limited to smartphones, tablets, laptops, and desktop computers. handheld devices, computer peripherals such printers and routers, mobile phones, electronic calculators, automotive control systems, and consoles. The ability to execute predetermined tasks independently is a defining feature of embedded computer systems. This means that design engineers may improve them in the ways described below. Like the basic 4-bit microcontrollers used to play a melody in a greeting card or a children's toy, there is less need for complete OS capabilities, such as memory management and scheduling numerous processes. This makes the product smaller and cheaper, allowing for more efficient mass production via economies of scale. An AV capture device, an AV player, a communicator, etc., might be a multi-function item. An example of a real-time restriction is an anti-lock braking system on a car, which requires the rapid release of the brakes in order to avoid the wheels locking up. This kind of system is common in embedded computer systems. It is possible to construct an all-pervasive system that functions like a Smart Grid by using models of multi-agent and autonomic systems. In order to comply with user regulations, such as reducing energy consumption, several devices may self-manage and work together. For instance, the system will choose which of several overlapping devices to turn on if it determines that some of them are unnecessary. There are a number of variables that affect how much it will cost to use energy, including the time of day, the energy rating (which differs from device to device), and the tariff. Customers may have a better understanding of their energy use and how to regulate it with the help of advanced utility consumption meters, which display consumption per unit-time and per device (Poslad, 2009)based upon three base designs: smart (mobile, wireless, service.

Decentralized Intelligence

The goal of decentralized artificial intelligence (also referred to as distributed AI) research is to provide distributed solutions to issues. Many consider it an antecedent of the studies that focused on software agents and (multi-) agent systems. Distributed problem solving was the umbrella term we used to describe this activity within the context of this book. To illustrate the point, consider a scenario where training neural networks with (extremely) large datasets would take a long time and a lot of resources on a single node. However, if several computing nodes are closely linked, they can divide up the "training work," resulting in lower operational time and better resource utilisation. So, distributed AI is connected to parallel computing in some manner. The most popular techniques and methodologies to parallel computing have been developed with a single stakeholder (e.g., a single user or organisation) in mind, who owns all of the data required to train the model. This is especially true in this context. You can't ignore the fact that things have been changing at a dizzying pace over the last few years. Some of the trends that contributed to the shifts are listed below.

To improve dependability, fault tolerance, and flexibility in ever-changing contexts, AI systems are moving towards distributed models.

- The widespread availability of high-powered portable devices equipped with a plethora of sensors, which provide data streams over which users may want some degree of control.
- The diminution in size and cost of sensors (and actuators), which may be "everywhere" and "anybody's"
- There are tiny, cheap, ML processors (like the NVIDIA Jetson Nano series devices) that can be embedded in almost any part of the IoT ecosystem.
- A proliferation of long-range and high-bandwidth wireless networks that link various devices and infrastructure components such as sensors, actuators, compute nodes, gateways, cloud(s), etc.
- Development, testing, and implementation of IoT ecosystems, touching almost every facet of daily life.
- Developments in techniques and their applications that can be used to different ML situations.

Consequently, methods that potentially enable coopetition are beginning to supersede the idea of a single data owner with all of the data kept in one place and utilised to train models to achieve their own (personal) objectives.

Human-Centric Networking

HCAL is the result of combining human-centered thinking with sophisticated algorithms based on AI. Technology has the potential to empower humans rather than supplant them, and our strategy will enhance that possibility. Developers and researchers used to put a lot of effort on creating AI algorithms and systems, with an emphasis on machine autonomy and performance measurement of algorithms. The new synthesis measures human performance and increases the value of user experience design, giving users and other stakeholders equal consideration. By prioritizing human values like rights, fairness, and dignity and by assisting human aims like self-efficacy, creativity, accountability, and social connections, researchers and developers of HCAI systems place a premium on meaningful human control. The rising trend to broaden the scope of technology-centered thinking to include human-centered goals that emphasise the good of society is reflected in this new synthesis. Since Al's 2017 Montreal Declaration for Responsible Development, interest in HCAI has increased. That proclamation urged commitment to privacy, autonomy, human well-being, and the development of a fair and just society. The Al4GOOD movement, Data Kind, and the IBM Watson Al XPRIZE Foundation share a passion for these human-centered objectives and want to use Al techniques, such ML, "to solve some of society's biggest challenges." The HCAI methodology, which uses rigorous design and assessment procedures to create high-impact research, is in line with this great dedication to social concerns.

HCAI approach that helps innovative designers prioritize people while developing AI solutions. Thermostats, lifts, self-cleaning ovens, smartphone cameras, and other commonplace equipment are only a few examples. Other examples include highly automated vehicles and pain management devices that patients operate themselves. Modern goals include extensive use of both human and automated control mechanisms. According to design metaphors, science and innovation are the two main tenets of Al research; but, in order to achieve their respective aims, researchers, developers, business executives, and policymakers will have to be resourceful in their pursuit of methods to combine the two for the benefit of end users (Shneiderman, 2022).

Resilience and Adaptability

The Intelligence of fast growth relationships of computers, robots, cyborgs, and artificial intelligence require the new AI nexus networks to be flexible and robust. For these networks to be even more entrenched in essential sectors such as health, finance, and transport, resilience and adaptability to the given environment become an immense priority. This section looks at the guidelines, issues and approaches to developing goals and outcomes for AI systems to be more resilient and adaptable systems.

The Pillars of Resilience

Resilience of AI nexus systems means their capability of coping and effectively responding to adverse instances which include but not limited to instances of cyber probing, hardware crashes, or disasters. Key components include:

1. **Fault Tolerance**: The ability to continue with operations at normal efficiency in spite of breakdowns of some of the system components. This is done through redundancy, failover and decentralizing capabilities throughout the topology structure of a network.

2. **Robustness:** Technique of testing that guarantees the ability of a system to operate and perform well in the presence of unexpected inputs or other poor data conditions. That is why they use such techniques as adversarial training and the use of noise-resistant algorithms.

3. **Self-Healing Mechanisms:** Implementing use of automated diagnostic and repair programs and procedures whereby various systems are capable of detecting problems and fixing them without outside help.

4. **Operational Continuity:** Contingency planning and developing back up standard procedures which enable organization to operate at other critical operations during a disruption.

Collaborative AI Ecosystems

AI ought to be human-centered, not the other way around, yet the emergence of the collaborative AI ecosystem changes our social and professional responsibilities by introducing new positions and necessitating specialised talents. Understanding these new dynamics is essential to our readiness in the changing workplace as we stand at the nexus of human knowledge and machine intelligence. Next, we will examine several new occupations and positions that are associated with the AI revolution. The position of an AI trainer stands out as perhaps crucial at the forefront. Training an AI model requires a thorough comprehension of data subtleties, biases, and real-world settings, in contrast to standard programming. Al trainers provide models sensitivity to these factors, guaranteeing that their results are consistent with moral principles and human values. This position requires a combination of technical skill and social awareness since it is less about coding expertise and more about providing wise counsel. In parallel, there is a discernible increase in need for Al ethicists as Al systems become more integrated into our everyday lives. Their accountability goes beyond just complying. To guarantee that Al is in accordance with human-centric ideals, Al ethicists provide recommendations while navigating the murky waters of moral quandaries and social effects. With a background in technology, sociology, and philosophy, their knowledge aids in the peaceful assimilation of Al systems into society without violating people's rights or fostering prejudices. As Al systems often engage in direct user engagement, the function of a human-Al interface designer gradually becomes more important. They specialize in designing user-friendly interfaces that make Al systems both practical and easy to use.

- We have explored a range of modes and roles that influence how we engage with artificial intelligence via human-Al cooperation. These revelations show how Al may contribute actively to our activities.

- Al and people are seen as having separate but complementary roles in parallel work, where they both work autonomously but ultimately work together to achieve a same goal. By combining human ingenuity with Al's computing prowess, this method has the potential to improve integration mechanisms and encourage the development of ethical standards in the future.

- Al is an auxiliary tool that improves human decision-making and creative exploration in constructive activity. As indicated by expectations of future Al systems, their contribution towards the direction of constructive labour is going to be more significant due to such factors such as empathy, context awareness as well as ethical compliance.

- Instead of a human-centric distribution of tasks, both human and Al participate in a common goal as a team. Human and Al agents engage, communicate, cooperate and change their actions based on the other's actions and needs. Thus, Al has the possibility to work on the empathy, contextual, and ethical aspects of alignment at work, which will make him even more useful for his team.

- Reciprocal cooperation best describes the interpersonal interaction between people and Al agents. At the core of this holistic and cyclical interaction is a generation of unpredictable outcomes, results, ideas and solutions. Any advancements in Al's ability to feel for, understand context and stay ethical will definitely influence synergy.

- Such hybrid models of cooperation also reflect the systems view of human-Al interactions where such cooperation is often context dependent. It is only when these models foster context-aware and adaptable Al systems, coupled with moral guides and industry standards that encourage productivity when working together.

Multiple Choice Questions (MCQs)

1. **Which of the following is a key application of AI-driven optimization in network management?**

   a. Static network configuration
   b. Dynamic routing and traffic management
   c. Manual resource allocation
   d. Data transmission via traditional protocols

2. **How does blockchain technology contribute to network security according to the CIA triad model?**

   a. By enhancing computational power
   b. By improving network traffic management
   c. By ensuring confidentiality, integrity, and availability of data
   d. By reducing the need for encryption methods

3. **Which quantum computing algorithm is considered one of the most important and well-known in the field?**

   a. Google's algorithm
   b. Shor's algorithm
   c. Grover's algorithm
   d. Bell's theorem

4. **Which of the following is NOT a role of IPCS in healthcare?**

   a. Telemedicine
   b. Emergency response and management
   c. Automated drug manufacturing
   d. Pervasive access to data

5. **What is the primary purpose of the context-aware computing layer in the healthcare system architecture based on monitoring?**

   a. To handle basic health data storage
   b. To make decisions based on environmental data like pollutant levels

    c. To manage cloud resources for efficient data access

    d. To process real-time healthcare images

6. **In the manufacturing industry, what benefit does integrate IIoT in inventory management provide?**

    a. Reduction of product quality

    b. Automation in tracking and packing inventory

    c. Minimization of data storage requirements

    d. Increased human effort in managing inventory

7. **What is the primary goal of "ubiquitous intelligence" in the context of AI Nexus Systems?**

    a. To minimize the energy consumption of devices

    b. To create intelligent systems that appear invisible to the user

    c. To focus on hardware improvements for AI systems

    d. To enhance centralized AI models for quicker training

8. **Which of the following is NOT a key component of resilience in AI Nexus systems?**

    a. Fault tolerance

    b. Real-time data processing

    c. Self-healing mechanisms

    d. Operational continuity

9. **What emerging profession focuses on ensuring AI systems align with human values and ethical standards?**

    a. AI Trainer

    b. AI Developer

    c. AI Ethicist

    d. AI Programmer

10. **Which of the following is an example of a system using decentralized intelligence?**

    a. A centralized database for AI model training
    b. A cloud server storing all data for AI processing
    c. Multiple distributed nodes working together to process large datasets
    d. A single computer managing all neural network computations

Answer

1	2	3	4	5	6	7	8	9	10
b	c	b	c	b	b	b	b	c	c

BIBLIOGRAPHY

Akhtar, M. W., Hassan, S. A., Ghaffar, R., Jung, H., Garg, S., & Hossain, M. S. (2020). The shift to 6G communications: vision and requirements. *Human-Centric Computing and Information Sciences*. https://doi.org/10.1186/s13673-020-00258-2

Ambedkar, B. (2015). *Fundamentals of Computer Networking (FCN)*.

Bhuyan, M. H., Bhattacharyya, D. K., & Kalita, J. K. (2017). *Network Traffic Anomaly Detection and Prevention: Concepts, Techniques, and Tools*. Springer International Publishing.

Celik, A., & Eltawil, A. M. (2024). At the Dawn of Generative AI Era: A Tutorial-cum-Survey on New Frontiers in 6G Wireless Intelligence. *IEEE Open Journal of the Communications Society*. https://doi.org/10.1109/OJCOMS.2024.3362271

Chio, C., & Freeman, D. (2018). *Machine Learning and Security: Protecting Systems with Data and Algorithms*. O'Reilly Media.

D, V. V. (2023). *The Intersection of Ai and Consumer Behavior: Predictive Models in Modern Marketing*. *8*(4), 2410–2424. https://www.mendeley.com/catalogue/e1d042c7-e098-3fce-bf79-7236659ff6a5/?utm_source=desktop&utm_medium=1.19.6&utm_campaign=open_catalog&userDocumentId=%7B41c321d5-81a1-4628-b5c2-1724042b9e8f%7D

Enoch Oluwademilade Sodiya, Uchenna Joseph Umoga, Alexander Obaigbena, Boma Sonimitiem Jacks, Ejike David Ugwuanyi, Andrew

Ifesinachi Daraojimba, & Oluwaseun Augustine Lottu. (2024). Current state and prospects of edge computing within the Internet of Things (IoT) ecosystem. *International Journal of Science and Research Archive.* https://doi.org/10.30574/ijsra.2024.11.1.0287

Esenogho, E., Djouani, K., & Kurien, A. M. (2022). Integrating Artificial Intelligence Internet of Things and 5G for Next-Generation Smartgrid: A Survey of Trends Challenges and Prospect. In *IEEE Access.* https://doi.org/10.1109/ACCESS.2022.3140595

Falcon, L. (2024). *Ethernet Performance in Computer Networks.* https://luckyfalconcomputers.com/ethernet-performance-in-computer-networks/

Fizza, M., & Shah, A. (2016). 5G Technology: An Overview of Applications, Prospects, Challenges and Beyond. *Proceedings of the IOARP International Conference on Communication and Networks (ICCN 2015)*, 18–19. http://ioarp.org/ioarp-admin-panel/upload/articles/1460357886_IDL-ICCN15-011.pdf

Fry, J. (2023). *Application of AI in Finance.* Jelvix. https://jelvix.com/blog/ai-in-finance

Gilbert, M. (2018). *Artificial Intelligence for Autonomous Networks.* CRC Press.

Glisic, S. G., & Lorenzo, B. (2022). Artificial Intelligence and Quantum Computing for Advanced Wireless Networks. In *Artificial Intelligence and Quantum Computing for Advanced Wireless Networks.* https://doi.org/10.1002/9781119790327

Goundar, S., Avanija, J., Sunitha, G., Madhavi, K. R., & Bhushan, S. B. (2021). Innovations in the industrial internet of things (IIoT) and smart factory. In *Innovations in the Industrial Internet of Things (IIoT) and Smart Factory.* https://doi.org/10.4018/978-1-7998-3375-8

Hammouch, Z., & Jamil, O. (2024). *Convergence of Antenna Technologies, Electronics, and AI.* IGI Global.

Juniper. (2024). *What is AI-Native Networking?* Juniper.

Kanthavel, R., & Dhaya, R. (2024). *AI for Large Scale Communication Networks*. IGI Global.

Kaur, J., & Khan, M. A. (2022). Sixth Generation (6G) Wireless Technology: An Overview, Vision, Challenges and Use Cases. *2022 IEEE Region 10 Symposium, TENSYMP 2022*. https://doi.org/10.1109/TENSYMP54529.2022.9864388

Kizza, J. M. (2016). *Ethics in Computing: A Concise Module*. Springer International Publishing.

Kong, N., Fosmire, M., & Branch, B. D. (2017). Developing library GIS services for humanities and social science: An action research approach. In *College and Research Libraries*. https://doi.org/10.5860/crl.78.4.413

Kumar, S., & Sharma, M. (2024). *A Comprehensive Review of Generative AI -From its Origins to Today and Beyond*. https://doi.org/10.13140/RG.2.2.19420.81281

Lin, X., & Lee, N. (2021). *5G and Beyond: Fundamentals and Standards*. Springer International Publishing.

Liu, M. Da, Chen, Z. N., Shi, Y. J., Tang, L. T., & Cao, D. (2021). Research Progress of Blockchain in Data Security. In *Jisuanji Xuebao/Chinese Journal of Computers*. https://doi.org/10.11897/SP.J.1016.2021.00001

Majumdar, A. K. (2019). Basics of Worldwide Broadband Wireless Access Independent of Terrestrial Limitations. In *Optical Wireless Communications for Broadband Global Internet Connectivity* (Issue 1, pp. 5–38). Elsevier. https://doi.org/10.1016/B978-0-12-813365-1.00002-3

Massimo, C., Mahmoodi, T., & Araniti, G. (2017). Software Defined Networking (SDN) and Network Function Virtualization (NFV) for C-RAN systems. In *5G Radio Access Networks: Centralized RAN,*

Cloud-RAN and Virtualization of Small Cells (Issue 1). CRC Press. https://doi.org/10.1201/9781315230870

Mori, D. (2022). *Manufacturing Basics: The transformation to Industry 4.0.* https://br.dmgmori.com/news-and-media/blog-and-stories/blog/manufacturing-basics-the-transformation-to-industry-4-0

Nand Kumar, E. al. (2023). Self-Healing Networks AI-Based Approaches for Fault Detection and Recovery. *Power System Technology.* https://doi.org/10.52783/pst.206

Neeraja, B., Yousoof, M., Badrinath, N., & Gupta, S. (2024). *Overview of CHATGPT -- AI TOOL.* RK Publication.

Negnevitsky, M. (2005). Artificial Intelligence: A Guide to Intelligent Systems, Second Edion. In *Polyhedron.*

Pandya, S. (2021). *A Study of the Recent Trends of AI in Healthcare, Immunology: Key Challenges, Domains, Applications, Datasets, and Future Directions.* https://www.researchgate.net/figure/A-representation-of-various-applications-of-AI-in-healthcare_fig3_355954943

Patil, G. (2023). *The Art of Balancing: A Comprehensive Guide to Load Balancers.* Medium.

Pawar, J. P., & Ingole, P. V. (2024). *Cognitive Radio Network with Artificial Intelligence.* OrangeBooks Publication.

Peng, J. (2021). *Machine Learning Techniques for Personalised Medicine Approaches in Immune-Mediated Chronic Inflammatory Diseases: Applications and Challenges.* https://www.researchgate.net/figure/The-main-types-of-machine-learning-Main-approaches-include-classification-and-regression_fig1_354960266

Perets, T. (2019). *INVESTIGATION OF WI-FI (ESP8266) MODULE AND APPLICATION TO AN AUDIO SIGNAL TRANSMISSION.* https://www.researchgate.net/figure/Wireless-Local-Area-Network-connected-to-the-internet_fig1_352750304

Poslad, S. (2009). Ubiquitous Computing: Smart Devices, Environments and Interactions. In *Ubiquitous Computing: Smart Devices, Environments and Interactions*. https://doi.org/10.1002/9780470779446

Rajendran, T., Shri Bharathi, S. V., Sridhar, S., & Anitha, T. (2023). A Study on Blockchain Technologies for Security and Privacy Applications in a Network. In *SSRG International Journal of Electronics and Communication Engineering*. https://doi.org/10.14445/23488549/IJECE-V10I6P107

Rehman, J. (2021). *What is local area network (LAN) in computer*. https://itrelease.com/2021/04/what-is-local-area-network-lan-in-computer/

Sangaiah, A. K., Shantharajah, S. P., & Theagarajan, P. (2019). Intelligent pervasive computing systems for smarter healthcare. In *Intelligent Pervasive Computing Systems for Smarter Healthcare*. https://doi.org/10.1002/9781119439004

Sharma, S., Sharma, K., & Sen, B. (2023). A Comprehensive Study on Natural Language Processing, It's Techniques and Advancements in Nepali Language. *Lecture Notes in Networks and Systems*. https://doi.org/10.1007/978-981-99-4284-8_13

Shneiderman, B. (2022). HUMAN-CENTERED AI. In *Human-Centered AI*. https://doi.org/10.1093/oso/9780192845290.001.0001

Singh, K. (2023). *Principles of Generative AI A Technical Introduction. 3*, 1–12.

Smith, R., Cubino, M., & McKeon, E. (2024). *The AI Revolution in Customer Service and Support: A Practical Guide to Impactful Deployment of AI to Best Serve Your Customers*. Pearson Education.

Sur, S. N., Imoize, A. L., Bhattacharya, A., Kandar, D., & Banerjee, J. S. (2024). *Artificial Intelligence for Wireless Communication Systems: Technology and Applications*. CRC Press.

Szeliski, R. (2011). Computer vision: algorithms and applications. *Choice Reviews Online*, *48*(09), 48-5140-48–5140. https://doi.org/10.5860/choice.48-5140

Tatineni, S. (2024). *AI-Driven DevOps Decision-Making: Transforming Software Development Workflows with Intelligent Systems*. Libertatem Media Private Limited.

Uchenna Joseph Umoga, Enoch Oluwademilade Sodiya, Ejike David Ugwuanyi, Boma Sonimitiem Jacks, Oluwaseun Augustine Lottu, Obinna Donald Daraojimba, & Alexander Obaigbena. (2024). Exploring the potential of AI-driven optimization in enhancing network performance and efficiency. *Magna Scientia Advanced Research and Reviews*. https://doi.org/10.30574/msarr.2024.10.1.0028

Vcelink. (2024). *Coaxial Cables: All You Need to Know*. https://www.vcelink.com/blogs/focus/coaxial-cables-all-you-need-to-know

Xiao, X. C., Zheng, X. W., Wei, Y., & Cui, X. C. (2020). A Virtual Network Resource Allocation Model Based on Dynamic Resource Pricing. *IEEE Access*. https://doi.org/10.1109/ACCESS.2020.3020944

Zhang, X. (2017). LTE Optimization Engineering Handbook. In *LTE Optimization Engineering Handbook*. https://doi.org/10.1002/9781119158981

ABOUT THE AUTHORS

Gopal Karamchand, known as GK, is an accomplished technology leader with an impressive career spanning over 15 years in cybersecurity, cloud computing, AI-driven security, and advanced data center architectures. As the Vice President of Information Security for a billion-dollar nonprofit organization, I have spearheaded innovative strategies, saving millions in costs while enhancing security and scalability.

My achievements include designing a global cloud network for HP, serving over 1 billion users annually, and pioneering Zero Trust security frameworks. My expertise extends to AI security, ethical hacking, quantum computing, and cloud networking, reflecting my relentless pursuit of technological innovation.

With multiple advanced degrees in Computer Science and Telecommunications, along with prestigious certifications like CCIE and Certified Storage Engineer, my academic and professional credentials underscore my mastery in IT infrastructure and cybersecurity.

As an inspiring thought leader and author, I channel my experiences into writing, sharing invaluable insights on technology's potential to transform industries and lives. Outside of my professional endeavors, I am a devoted family man and an advocate for Ayurveda and holistic wellness. I reside in Houston, Texas, with my wife, Deepika, and our two children, Anvi and Krish.

Satish Chitimoju, born in Visakhapatnam, Andhra Pradesh, India, embarked on his journey in the serene village of PB Palli. With a degree in Electronics and Communication and an early career in banking technology, Satish gained invaluable international exposure working onsite at EastWest Bank's main branch. His passion for innovation led him to Bangalore's tech ecosystem, where he excelled with companies like Accenture, Infosys, and SmartBear, before contributing to Hyderabad's dynamic startup scene.

With 14 years of experience, Satish has mastered full-stack development, big data applications, AI/ML solutions, and AIOps/MLOps. His expertise spans industries such as banking, healthcare, automotive, and travel, delivering impactful solutions on-premises and across cloud platforms like AWS, Azure, and GCP. A certified AWS and Azure professional, he has been recognized as a guest speaker in Houston, Texas, for presenting AI-driven logistics innovations.

As a leader and thought advocate, Satish inspires startups, empowers communities, and shares insights through writing, showcasing the transformative power of technology. His commitment to innovation, mentorship, and societal progress defines him as a goal-oriented professional and a catalyst for change.